Caribbean Wild Plants
and their Uses

An Illustrated Guide to some Medicinal and
Wild Ornamental Plants of the West Indies

Penelope N. Honychurch

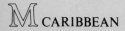 CARIBBEAN

Original edition 1980
This edition 1986
Reprinted 1987, 1989

Published by *Macmillan Publishers Ltd*
London and Basingstoke
Associated companies and representatives in Accra,
Auckland, Delhi, Dublin, Gaborone, Hamburg, Harare,
Hong Kong, Kuala Lumpur, Lagos, Manzini, Melbourne,
Mexico City, Nairobi, New York, Singapore, Tokyo

ISBN 0-333-40911-6

Printed in Hong Kong

British Library Cataloguing in Publication Data
Honychurch, Penelope N.
    Caribbean wild plants and their uses: an
    illustrated guide to some medicinal and wild
    ornamental plants of the West Indies.
    1. Herbs — Caribbean Area — Therapeutic use
    I. Title
    615'.321   RM666.H33
ISBN 0-333-40911-6

Front cover *Jacquemontia pentantha* (Jacq.) G. Don

Dedicated to Iris.

*This is also for Lawrence Halprin and those
who gave me plants and information.*

Contents

Preface

Sixteen years ago I started making drawings and notes on wild plants of the Caribbean used as remedies, and the subject soon became of absorbing interest — and grew somewhat haphazardly as I started to include wild plants which had no medicinal value but which seemed attractive or unusual.

As time has gone by, I have realized that many of the ways in which plants were used are dying out, as customs change and the patterns of West Indian life become more like those of industrialized countries. My concern also extends to wild plants which will disappear from these islands unless there is a greater awareness of the need to preserve natural resources and to initiate sensible conservation policies. I hope that perhaps this work might stimulate similar endeavours, as there is still so much to be learned from the plant world.

I am deeply grateful to Mr. Douglas Taylor for his help. Any references to Carib use of plants were supplied by his work (xix). I am indebted to Dr. R. A. Howard who took the time to identify many of the drawings included, and to Dr. C. D. Adams, who has also helped with drawings and specimens. I am also indebted to Mr. E. G. B. Gooding, who kindly helped me with some plants I could not identify and with some of my descriptions, and to Dr. D. H. Nicolson, who helped in plant identification as well.

The final shape of the medicinal sections of this work has been largely due to Mrs. Iris Bannochie, who supplied most of the medicinal information from Barbados by lending me her notes. Without her encouragement this work might never have reached this stage.

I owe a great debt to the people in Dominica and in other islands who showed or gave me plants they used — their contributions have been extensive, particularly that of Mr. Michel Grandguillotte in Marie Galante. To my editors Dorothy Cameron and Kay Wood I owe a special thanks, and to Deborah Dunn, who, as a botanist, was helpful in correcting the proof, and to Carol Bunting who typed the many pages of notes.

In the naming of the plants I have relied mainly on Dr. R. A. Howard, Dr. C. D. Adams, Dr. D. H. Nicolson and Mr. J. Fournet. All reference sources are listed at the back of the book.

Introduction

In this book are described some of the wild plants in the Caribbean which are used as remedies for illnesses or have folklore associations. It also includes wild plants which have no medicinal value but are often found in the same family, and are, in some instances, scarce. Most of the plants have been illustrated to make identification easier. The drawings show plants at approximately one-sixth their natural size. An asterisk beside the Latin name in the text indicates that an illustration is included.

The usual botanical order is not followed in arranging the plants described. In each of the three parts, the families have been arranged alphabetically, as have the genera within them.

Local names as well as the Latin names for plants have been given wherever possible. These names may differ, even in different parts of the same island. Thus 'Bonavis' in Giraudel may be 'Boucousou' in Pennville (Dominica), or 'Ma Bizou' in Citronniere may be 'Mal Visou' in Sylvania (Dominica). Sometimes the French or patois term for a plant (given throughout in inverted commas) is a direct reflection of the original botanical name.

There are roman numerals given in the body of the text — these refer to my sources of information which are listed in the references at the end of the book.

Three separate indexes are provided to French or patois, English, and scientific names.

My only claim to direct medical connections is a short list of medicinal compounds, together with the plants in which they are found (pp. 137–138).

In a useful summary, the Caribbean wild plants are classified under the ailments for which they were or may be used as the remedy (pp. 139–145).

Lists of plants which may be used for fodder and those which serve as a source of nectar for honeybees are included in the book.

Botanical terms which may be unfamiliar to the reader are defined in the Glossary.

A note on the medicinal use of plants

General

Knowledge of herbs and their uses in the Caribbean originated with slaves brought from Africa and with the Caribs who then inhabited many of the islands. This knowledge was handed down from generation to generation, changing as new uses were discovered. So, although some plants in the Caribbean are the same as those in West Africa, uses have varied as local habits took different forms and as different people settled on the islands. For example, uses of plants in the Saintes by descendants of Bretons might correspond to uses of plants by the French in St. Thomas. These same plants might be used quite differently in Trinidad, Barbados or Dominica.

Dominica has served as the main point of reference for this work, and in Dominica much knowledge of herbal remedies has remained with the Caribs, although it is hard to say what they themselves learned from the arrival of new cultures.

Plant remedies for illnesses or complaints are still used, but some have changed or been discontinued. Others are known only by hearsay. In such cases, I have described their use in the **past tense**. Undoubtedly some of these old remedies are no longer used because there is a wider distribution of modern medicine and easier access to professional medical help. Nowadays, for example, epsom salts is used more often than a plant such as *Cuscuta americana*. Conditions in each island vary widely, but certainly in most of the islands, older people use many herbal remedies, especially those people living in remote areas.

These remedies are becoming milder, because it is now realized how **poisonous** some plants can be. Previously, a strong all-round bush tea made up of different plants, some of which were poisonous, was given in the hope that at least one of the plants in it would cure the complaint. Unfortunately, it sometimes had the opposite effect, and resulted in severe inflammations and even in death. Children were sometimes victims of bush tea poisoning (viii). They were often given soursop tea (*Annona muricata*) to make them sleep, and in cases of malnutrition, were given worm teas which were much too strong (vi). *Crotalaria fulva* has been cited as a factor in causing what is known as veno-occlusive disease of the liver (V.O.D.), and 30% of all cases of liver cirrhoses at the University Hospital of Jamaica were attributed to V.O.D. (xxviii).

Apart from remedial uses, several plants still have associations with obeah, such as *Thevetia neriifolia, Abrus precatorius, Hippomane mancinella*, or *Nerium oleander*. Often these could be cut only at certain times of the moon, and some were sometimes soaked overnight.

Descriptions of illnesses throughout the islands tend to be vague, and while a 'fresh cold' is precise, 'bad feels' is more challenging. Thus one might get a tea designed to cure cough or cold, and wind or worms at the same time. A popular illness description in Dominica is 'a pain in the waist'.

It is hard to know what quantities were or are used of a particular plant. In my own experience I have been given a handful of different herbs to be boiled into a 'red' tea to relieve a bad cold. At another time I have been told to use only seven or nine leaves of a certain plant to be taken as a purge. The strength of these plants varies a great deal. In resinous plants, older leaves may be better than young shoots, and in some cases, dried leaves are stronger than fresh.

These curative practices may eventually die out as scientific research continues and customs change, but I hope that the beneficial effects to be derived from plants will continue to be studied and utilized. In Marie Galante it is said 'tout hazié sé rimèd' or 'every bush has a remedy'.

Baths

While it is generally felt that there are plant remedies for every illness, it is interesting that there seems to be no conflict between the use of baths as protection against evil spells and traditional religious convictions (xxv). Such baths are taken for good luck, against 'piai', or 'l'amarage', that is, protection against malignant spells and mischievous forces, and for ritual cleansing. Taken in many forms, they are primarily protective. Most are taken only at certain times of the moon, or the year. Some are best taken on the first Friday of the new year or the quarter of the year, and before or after the full moon (x). These baths, unlike those which are said to remedy specific illnesses, will often consist of many plants such as: Lavande, Z'Herbe Grasse, Tabac Zombie, Caapi Doux, Caapi Marron, Kudjuruk, Pois Angol, Simen Contra, Grain en bas Feuille, and Bouton Blanc (x). Such baths are considered charms against problems that may confront the country dweller during his daily routine.

Fishermen (xxv) on the other hand, will often use baths for their canoes. These will ensure that their boats are safeguarded against malignant spells, bad weather, a poor catch, or any other disaster which might be natural, but which might also be the result of directed, or imagined, human hostility.

Many remedial baths are for childbirth, or are used immediately after, or during weaning, and are taken by the mother in the hope of producing whiter milk, or to ease lactation.

Others are said to cleanse the skin, to refresh, to relieve prickly heat, and to heal wounds and sores. Some are taken for poisoning. In the summary of medicinal plants, there is a list of the various kinds of baths taken.

Poultices

Wounds, bruises, sores, sprains or strains are generally remedied by poultices. For open wounds or sores, poultices are applied where the leaves are heated or crushed. Sometimes salt, sugar, sulphur or rum is used, as well as beer or honey, and mixed in with the leaves, bark, root or fruit.

Teas

Teas are commonly used for headaches, colds, fevers, chills, stomach troubles, and problems related to pregnancy. The Caribs used teas for charms, spells, and for their dogs to excite them to hunt. They carried leaves for good luck, and they avoided some plants because these were associated with spirits. Certain plants are still grown in gardens to guard against bad fortune or malignant spells. Many of these same plants are used in teas.

It is believed that, in order to be effective, the temperature of the tea must be correct for the complaint. Some plants can be used either hot or cold, but the temperature depends upon the time of day, weather conditions, and the illness itself.

Hot Teas (Decoctions) are usually taken for gripe, gas, purges, rheumatism, bad colds, fevers, chills, or to induce menstruation. These are sometimes bitter, and they are usually taken for several consecutive days.

Cooling Teas (Infusions) are, contrary to hot teas, believed to refresh and to relieve tension, but like hot teas, are often taken for several days at a time. They are used mainly to cleanse the blood, to cool the body, and for inflammation. They may also be taken after a purge.

Both hot and cold teas are sometimes taken with salt, sugar, rum, beer, vinegar or wine, depending upon the recipe.

Certain fruits have the same hot/cold associations. It is believed unhealthy to eat bananas, for example, after a certain time of day. Eating cold fruit in the hot sun or when over-fatigued is not considered healthy, nor is it considered wise to eat a 'cold' fruit after a hot tea.

Part 1

DICOTYLEDONS

ACANTHACEAE

Justicia pectoralis (Jacq.) Murr.*

Local names: Garden Balsam (Barbados); Carpenter's Grass (Jamaica); 'Z'herbe Charpentier' (Dominica and Martinique).

Found from Mexico to the northern part of S. America and in the West Indies. In neglected areas (i).

This herb has a trailing habit and is about 30 cm high. The leaves are lanceolate and have a reddish tinge. The flowers are small and pink with white lines and are arranged in spikes. The stems are slightly swollen above the nodes.

Medicinal uses Used as a tea for colds. When mixed and boiled with lemon grass, sugar, water, and hibiscus flowers it makes a pleasant tasting cough syrup. It has also been used as a tea after an injury or fall. The leaves and stem can be bruised and soaked in rum or water as a poultice for cuts or bruises (iv). When crushed they can be applied to sores for quick healing (x). It is used hot in Martinique for cold chills, and in Marie Galante it is used for stomach-aches (xii).

Odontonema nitidum (Jacq.) Kuntze*

Found in secondary forests on the borders, and in waste places at upper elevations, it is a native of the Lesser Antilles and Trinidad (i).

This shrub will grow up to 60–150 cm. The leaves are lanceolate, entire, and soft textured. The bright mauve flowers are smooth and tubular. They occur in spikes. There are several spikes on a stem.

Pachystachys coccinea (Aubl.) Nees*

Local names: 'Chandalier,' 'Plumet d'Officier' (Martinique).

Native of Guyana (i), it has become naturalized in the West Indies, and was introduced in Martinique, but apparently it is not found in Guadeloupe (xi). It grows most commonly in the southern part of Dominica. It is found by roadsides and in damp places in middle to upper elevations.

This plant is about 1 m high. It is conspicuous for its inflorescence. The stem is fleshy, and the elliptical leaves are about 15 cm long. The inflorescence is a brilliant scarlet with yellow stamens.

It was probably at one time introduced to Dominica as an ornamental, but it is now found only as an escape.

Medicinal uses The Caribs used the leaves as a tea for headaches (xix)

Ruellia tuberosa L.

Local names: Minnie Root, Duppy Gun (Barbados); 'Patate à Chandelier' (Marie Galante).

Found in St. Thomas, the Saintes, — from the southern part of the United States to the West Indies. It grows in dry places, pastures and other flat areas.

A low plant. The leaves are light-green, elliptical and simple. The blue flowers are about 2.5–5 cm long and funnel-shaped, with five petals. They occurs in clusters. When the pod, which is a narrow capsule, is handled it explodes. The roots are swollen and slightly elongated.

Medicinal uses The tubers and leaves are used with Clammy Cherry (*Cordia obliqua*) and soursop (*Annona muricata*) for blood disorders (vii). The roots and leaves are used for inflammation of the intestines, colds, coughs, and as a cooling drink (viii, xvi). It can cause haematuria if used to excess (viii). It was also used for cystitis and enteritis.

2

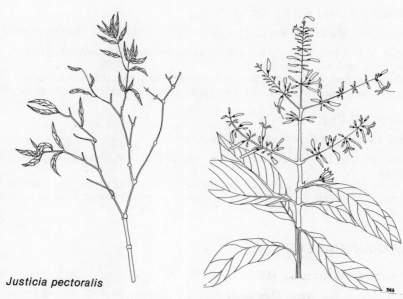

Justicia pectoralis

Odontonema nitidum

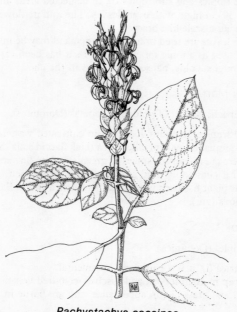

Pachystachys coccinea

3

ACANTHACEAE (continued)

Thunbergia alata Boj ex. Sims*

Local names: Black-Eyed Susan; 'Fleur Jaune Savane' (Martinique and Guadeloupe).

Native of East and South Africa, and found in Dominica on roadways, pastures, and overgrown gardens.
   A trailing plant with slender stems, it has winged leaf stalks, and the leaves are cordate. The flowers are deep yellow with a deep purple centre and occur profusely on the plant.

Thunbergia fragrans Roxb.*

Local name: White Nightshade (Jamaica).

Native of Asia, now widespread (i), this vine is found on roadways and overgrown gardens.
   Trailing in habit, this plant has slender stems, and cordate leaves. The stalks are not winged. The flowers are white with a light-green centre. Flowering is not as profuse as *T. alata*.

AMARANTHACEAE

Achyranthes indica (L.) Mill*

Local names: Colic Weed, Man Better Man (St. Thomas); Hug-me-close (Barbados).

Generally found in the tropics and subtropics (i), in neglected areas and roadsides.
   An erect herb about 30 cm high, with ovate leaves. The minute flowers are whitish, in spikes, each flower having a scale-like bract.
Medicinal uses   The leaves are used in teas for fevers, and may be mixed with *Mimosa pudica*. The tea is also used as a tonic for 'tired blood'. It has been variously used as an aphrodisiac (viii), for coughs, colds, nausea, and pains in the chest.

Amaranthus dubius Mart.*

Local names: Wild Spinach; 'Bhaji' (Trinidad); 'Zepina' (Dominica).

Found generally in the tropics, in sunny locations in cultivated wastelands.
   This herb is an erect annual 60–90 cm tall, with a thick flaccid stalk. Some of the leaves may be toothed and there are numerous spikes of greenish-white flowers. It seeds quickly and grows rapidly, and is commonly used as a substitute for spinach.
Medicinal uses   This plant has been used in parts of Africa and in Barbados as a poultice for abscesses and boils (viii).

Iresine herbstii Hook.*

Local name: 'Zizier Poule' (Dominica).

Found in tropical America (i), it is a cultivated ornamental.
   A low shrub with deep maroon leaves, and fuschsia-coloured venation. The greenish-white flowers are small and grass-like in appearance, and are borne in racemes.

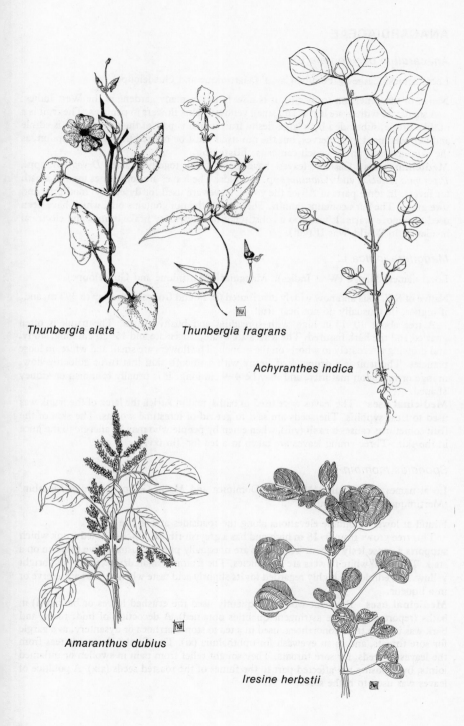

Thunbergia alata

Thunbergia fragrans

Achyranthes indica

Amaranthus dubius

Iresine herbstii

ANACARDIACEAE

Anacardium occidentale L.

Local names: Cashew, 'Pomme-Cajou' (Martinique and Guadeloupe).

Native of the American tropics (i), it is now found in many gardens in the West Indies.

A small tree with ovate leaves and small yellow to pink flowers in panicles. The fruit is a nutlike drupe, subtended by a large fleshy fruit which is part of the ovary. This is edible and has been used in preserves, but the nut itself must be roasted before consumption, as the skin gives off an oil which can cause **blistering**.

Medicinal uses Cashew leaves have been used together with *Dryopteris* spp., *Peperomia pellucida* and *Adiantum* spp. for colds. The leaves are sometimes used as a bath for fevers. In some parts of Africa the young leaves are used for dysentery, toothache and sore gums. The bark contains tannin. The shell of the nut contains oils, which have been used to remove warts. It was also a constituent of heavy-duty brake linings and electrical insulation in World War II (viii).

Mangifera indica L.

Local names: Mango (West Indies); 'Manguier' (Martinique and Guadeloupe).

Native of S.E. Asia and now widely distributed (i). Found from sea level up to 393 m, and, if higher, they usually do not bear fruit.

A tree about 10–15 m high with spreading and bushy crown. The trunk is often gnarled and the bark fissured. The leaves are ovate, lanceolate and 15–20 cm long, shiny, and growing alternately in whorls on the branch. The flowers are small and white, in loose panicles. The fruit is a drupe. It is fleshy with a smooth skin that turns golden-yellow, orange or red upon maturity, and may be 4–9 cm long. It is usually roundish or kidney shaped.

Medicinal uses The leaves were used in baths, and in Africa the latex of the trunk was used to treat syphilis. The seeds are said to get rid of intestinal worms. The skin of the fruit sometimes causes a **rash** (viii), when eaten by people who may be allergic to the juice in the skin. Three young leaves are taken in a tea for 'flu (xvi).

Spondias mombin L.

Local names: Hog plum, 'Mombin' (Dominica and Marie Galante); 'Prune Mombin' (Martinique and Guadeloupe).

Found at lower to middle elevations along the roadsides and in the fields.

This tree grows from 8–18 m high and has a greyish-ribbed, thick-skinned trunk which supports a dense leafy crown. The leaves are unequally pinnate, usually five to seven on a stalk. The small white flowers are in panicles. The fruit is a small drupe, fleshy and bright yellow. It is edible and highly regarded for its slightly acid taste when used as a preserve or in a liqueur.

Medicinal uses Pregnant women frequently used the crushed leaves or bark (xvi) in baths (reportedly for the astringent qualities obtained). A decoction of bud, roots, and bark was used to treat gonorrhoea, used in a tea to stop diarrhoea or dysentery, as a gargle for sore throats, and as an eyewash for ophthalmia (xi). The Caribs also made teas from the leaves or seeds, for sore throats. They sought relief from pain in swollen or inflamed joints, by exposing the affected part to the fumes of the roasted seeds (xix). A poultice of leaves was used to bathe sores.

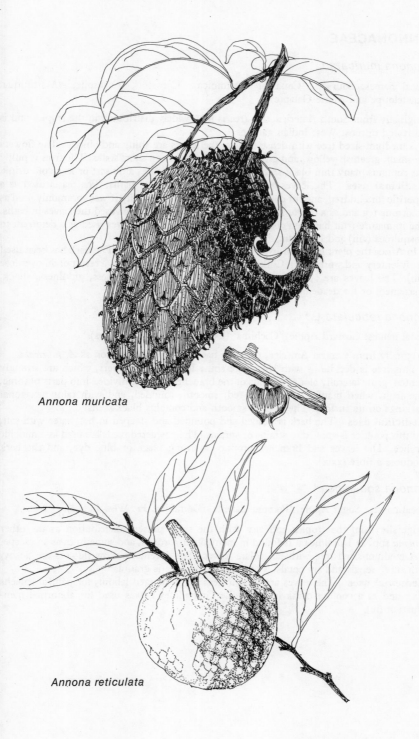

Annona muricata

Annona reticulata

ANNONACEAE

Annona muricata L.*

Local names: Soursop, 'Couassol' (Dominica); 'Corossol', 'Corosolier' (Martinique, Guadeloupe and Marie Galante).

Originally from South America, this tree is now widely distributed in the tropics and is cultivated in most West Indian gardens.

A medium-sized tree with dense foliage, the leaves are shiny and oblong. The flowers are small, greenish-yellow, and grow from the branches on short stalks. The fruit is pulpy and contains many thin black seeds. It is commonly used as a dessert or as a cold drink.
Medicinal uses The leaves are widely used in teas and baths. The tea is used as a soporific for children, and for colds and fevers. In Barbados (vii), it was commonly used as a morning tea and as a tea for heat and rashes. Women in labour used the leaves in baths. The immature fruit has been cooked as a vegetable, and the leaves used as a deterrent to mosquitoes (viii) and lice (xxix).

In Africa the plant has been used for coughs, colds and fevers. The bark has been used for dysentery and worms (iv). Children have been known to **die** as a result of an overdose (viii). The leaves are rubbed against the face in cases of dizziness, giddiness, shock, amazement or for deadening of pain (xvi).

Annona reticulata L.*

Local names: Custard Apple, 'Cachima' (Dominica, Fr. West Indies).

Originally from Central America, this tree has the same distribution as *A. muricata*.

This tree is deciduous with lanceolate shiny leaves. The flowers, which are strongly scented, grow laterally along the ends of the branches, and are divided into parts of three. The fruit, when mature, is greenish-red, smooth, rounded, and with faint hexagonal markings on its surface. The pulp is smooth and contains black seeds.
Medicinal uses The bark is pulped and pounded and steeped in hot water with salt, and this poultice is applied to strains or sprains. The powdered seeds are used as a miticide for lice. The leaves and branches produce a strong black or blue dye, and the bark produces a fibre (xxix).

Annona squamosa L.*

Local names: Sugar Apple, 'Pomme Canelle' (Dominica, Fr. West Indies).

Originally from the tropics and also with the same present distribution as the other *Annona* spp. In Dominica it is found in many older gardens, and sometimes as an escape.

A medium-sized tree with lanceolate leaves. The greenish-yellow flowers are fleshy, with three sepals and six petals. The pulp of the fruit is granular.
Medicinal uses The leaves are used in teas for prolonged labour, for painful spleens (xix), and as a cooling drink (viii). In the Grenadines it was used for abnormal menstruation (iv).

ANNONACEAE (continued)

Cananga odorata (Lam.) Hook f. & Thom.

Local name: 'Ylang Ylang'.

A native of Malaya (i). Cultivated in Dominica at both coastal and upper elevations, and sometimes cultivated in the West Indies.

The tree grows up to 20 m tall, and has a spreading rangy habit. The trunk of the tree is smooth and light grey, and the lateral slender branches radiate from the trunk at approximate right angles. The leaves are lanceolate, about 15 cm long and noticeably veined. The tree is cultivated primarily for its scented flowers. These are greenish-yellow, with three small sepals, and six ribbon-like petals which have a small red dot at their base. There are many small stamens and a round pistil. The fruits are in clusters of small oval navy-blue berries.

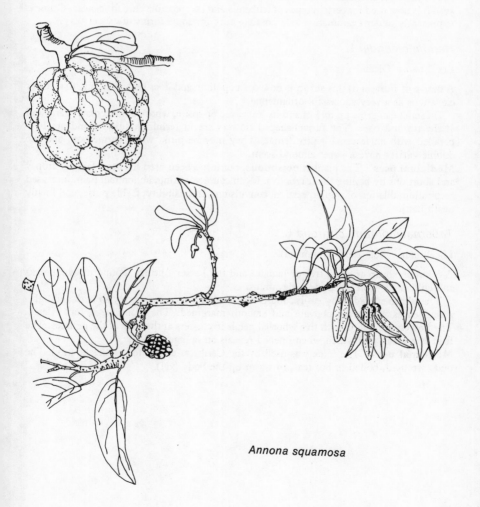

Annona squamosa

APOCYNACEAE

Catharanthus roseus (L.) G. Don*
(syn. Vinca rosea L.)

Local names: Periwinkle, Old Maid (Barbados); 'Caca Poule' (Dominica); 'Pervenche Blanche' (Marie Galante).

From Madagascar, it is now grown throughout the tropics as an ornamental. It may also be found in neglected areas as an escape.

This is a low herbaceous plant with slightly woody stems and a straggling habit. It grows to about 60 cm. The leaves are elliptical, oblong and shiny. The flowers occur on the ends of the branches in clusters. They are pink with a maroon centre. *C. occellatus* is white with a maroon centre, and *C. albus* is white with a greenish centre.

Medicinal uses   The leaves have been used as a tea for diabetes by the Caribs (xvi, xix) and in Jamaica, Grenadines and Africa. *C. albus* has been used for high blood pressure (iv, xvi). It is now used for certain types of leukemia and as a possible cure for cancer. *C. roseus* is presently under examination as a possible cure for some forms of cancer (xxvi).

Nerium oleander L.

Local name: Oleander.

A native of Eurasia (i) this shrub is now cosmopolitan and does best in sunny, dry, lower elevations as a very successful ornamental.

Oleander can grow up to 3 m and its leaves are opposite, whorled, lanceolate, narrowed at the tip and base. The funnel-shaped flowers are in terminal clusters, and are five-petalled with an internal centre fringe. They may be pink, white or maroon, and the double variety have a sweet almond scent.

Medicinal uses   The plant is **poisonous,** but it has been used for heart failure, dropsy and abortions by boiling the leaves (iv). It can cause vomiting, abdominal pain, increased respiration, dilation of pupils, vertigo, convulsions, insensibility, feeble pulse, and finally death (xxi).

Tabernaemontana citrifolia L.*

Local name: 'Bois Lait' (Dominica).

A common shrub or small tree of Jamaica and the Lesser Antilles. Found in windbreaks and cultivated wastes at middle elevations.

This shrub spreads very easily, and is noticeable for its smooth greyish bark and light-green leaves which are obovate and smooth margined. The plant has a milky latex. The flowers are white, with five whorled petals in clusters at the axils of the petioles. The fruit is a green pod which when opened reveals an orange rind.

Medicinal uses   The latex was used by the Caribs as a cure for toothache (xix). The buds are used, boiled in hot teas, to warm up the body (xvi).

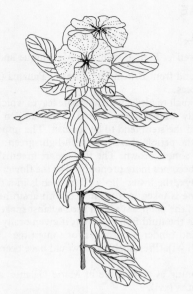

Catharanthus roseus

Tabernaemontana citrifolia

ARISTOLOCHIACEAE

Aristolochia trilobata L.*

Local names: 'Trèfle', 'Tweff', 'Tref' (Dominica, Martinique and Guadeloupe).

Found in the West Indies and from British Honduras to Panama (i), in secondary forests and lower bushy coastal areas.

A vine which climbs by small tendrils, with alternate leaves which are deeply lobed into three parts and are slightly leathery, yet flaccid. They are on long slightly curving petioles, and both these and the stem tend to be reddish. The upper surface of the leaf is glossy, and at the base of the petiole there is a small light-green cup-shaped bract. The young leaves are deep red, almost brown. The flowers are greenish with a curious deep-red netted venation which becomes more pronounced as the flowers mature. They consist of a tubular, pendulous calyx, the lower 'lip' of which ends in a flat ribbon-like tendril.

Medicinal uses This vine is taken in a tea mixed with absinthe for gripe or stomach pains. Evidently, all parts of the plant have been used against snake bites, both as teas and as poultices. It was said that the fluid from the root, if given orally in two or three drops, would intoxicate a snake and render it harmless for some time. A stronger dose would cause convulsions and death (xi). (Presumably this would have been done before the snake attacked.)

It is also taken for poisoning as a vomative in Marie Galante, and three, five or nine leaves are used in ritual baths (xvi).

There is conjecture that one of the many types of leaf used in baths against bad luck or bad spells, called 'piai', was a species of *Aristolochia*, taken by the Caribs (xix).

ASCLEPIADACEAE

Asclepias curassavica L.*

Local names: Kittie McWanie, Blood Flower (St. Thomas); 'Matac' (Dominica).

Commonly found in the West Indies, in pastures and neglected gardens from the coast to about 328 m.

A slender erect weed, it has bright red and yellow flowers borne terminally in cymes. The pods (not shown) are narrow and filled with brown seeds which have silky hairs. The stem and flower stalk contain a milky latex.

Medicinal uses The plant is **toxic**. Small livestock have died from grazing on it. The roots were pounded into an infusion as an emetic which could also be used to poison fish. A poultice was used to treat ringworm, and in Jamaica to stop bleeding (iv). The Caribs (xix) considered the root to be a good febrifuge, and in Africa it was used for intestinal troubles in children (iv).

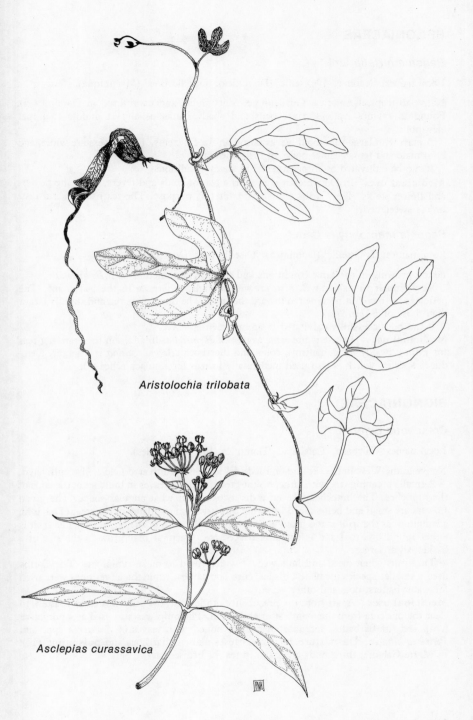

Aristolochia trilobata

Asclepias curassavica

BEGONIACEAE

Begonia hirtella Link*

Local names: Begonia; 'L'Oiseille' (Dominica); 'Oiseille Bois' (Martinique).

Native to tropical America, Guadeloupe, Martinique, and naturalized in Dominica (i). Found in ravines, valleys, river beds and shady banks usually at middle to upper elevations.

A herb with large rhomboid leaves borne on flaccid stems. The flowers are white, and both male and female flowers are on the same stalk.

It can be cultivated as an ornamental in shady conditions.

Medicinal uses In Dominica it is used as a tea for colds and fevers, using the flowers and flower stalks. The Caribs also used it for this purpose. The tea has an acrid taste unless sweetened.

Begonia macrophylla Lam.*

Local names: 'L'Oiseille' (Dominica); 'Oiseille Bois' (Martinique).

Found growing in the same conditions and often in conjunction with *B. hirtella*.

Very similar in habit to *B. hirtella*, with a slight difference in the leaf shape. The petioles are green but become red toward the stem. The flowers are red and slightly larger than those of *B. hirtella*.

It can be successfully culvitated as an ornamental.

Medicinal uses Its use is the same as that of *B. hirtella*. In addition there are hot teas for gripe, 'fwaidy' (rheumatism), colds and diarrhoea. *Begonia nitida* is a larger, introduced species, which is also used medicinally in teas for stomach-aches (x).

BIGNONIACEAE

Crescentia cujete L.*

Local names: Calabash, 'Calebasse' (Dominica, Fr. West Indies).

Native to the West Indies. Found on roadsides and once-cultivated fields. Also cultivated.

A small spreading tree with large, bright-green, obovate leaves in bunches at the ends of the branches. The branches take on a distinctive, somewhat angular shape. The green flowers are small and bell-shaped, growing directly from the trunk of the tree. (This is an adaptation, as the fruit is too heavy to be borne on the ends of the branches.) The fruit is either round or ovoid and is smooth-surfaced and green. It may grow to the size of a basketball or larger.

The fruit, when dried and hollowed, is widely used to make containers. The Caribs used a smaller species for plates, dishes, cups and rattles, while the larger ones were used for water bailers, pots and other vessels.

Medicinal uses Apart from the practicality of the fruit, the pulp was eaten roasted, to clear the placenta from the womb after childbirth. The pulp was also used as a purgative (viii), and, in Barbados, for abortions when boiled with leaves of *Swietenia* spp. and *Petiveria alliacea*. The mixture, however, causes nausea, diarrhoea and **poisoning** (viii). In Marie Galante, three buds are taken in tea for bruises (xvi).

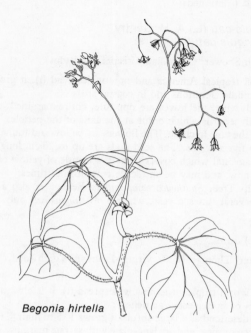

Begonia hirtella

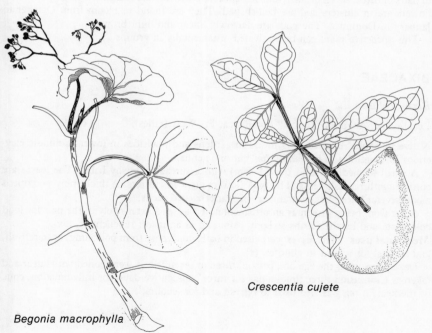

Crescentia cujete

Begonia macrophylla

BIGNONIACEAE (continued)

Macfadyena unguis-cati (L.) A. H. Gentry*
(syn. Doxantha unguis-cati (L.) Meirs)

Local names: Golden Shower; Cat's Claw Creeper (Barbados).

Native of continental tropical America and now widespread (i). It grows in secondary forests at lower to middle elevations and in coastal areas.

A spreading woody vine, the leaves are opposite, entire-margined, and ovate, about 7.5–10 cm long, with tendrils which occur at the axils of the petioles. The tendrils are clawed, which helps the vine to climb. The flowers are yellow and funnel-shaped, and the petals are lobed into five sections. The seed pods are up to 20 cm long and linear.

A spectacular ornamental which covers trees in a mantle of yellow blossoms, it is not generally cultivated now, and may be an escape from former times.

The African Tulip Tree, Spathodea campanulata Beauv., is also a member of this family. It flowers several times a year, whereas M. unguis-cati only blossoms during May.

Tecoma stans (L.) Juss.*

Local names: Ginger Thomas (U.S. Virgin Islands); Torchwood, Christmas Hope (Barbados).

Native of the New World tropics it is now widespread (i). It is always found in coastal areas along roadsides or in neglected gardens.

A small tree, also described as a large shrub, it grows 1–5 m high. The leaflets may be in pairs of three, seven or nine, and are lanceolate with serrate margins. The bright yellow flowers are in clusters and are bell-shaped. They are found in bloom from October to January in Dominica. The pods are narrowly linear and light brown.

This attractive plant can be cultivated ornamentally in groups.

BIXACEAE

Bixa orellana L.*

Local name: Annatto, 'Roucou' (Dominica, Fr. West Indies).

Native to South America, it is found in gardens and roadsides in lower to middle elevations in Dominica. This family has but one genus.

A small to medium-sized tree (3–5 m) with a spreading bushy habit. The leaves are cordate, with reddish veins. The pink flowers grow in clusters. A thick spiny pod opens along several lines to show numbers of scarlet-coated seeds.

Apart from being grown as an ornamental, it was used extensively in the past for food colouring, and by the Caribs as body paint and as an insect repellent (x, xix).

Medicinal uses The leaves were used in baths to cleanse from poisoning, or to refresh, and in teas, for worms in children (x).

Depending upon the age of a person, three to seven leaves were boiled with sugar and taken as an infusion three times a day for three consecutive days, for inflammation, colic or the heat. A tea was also used to refresh and for 'cooling'.

Macfadyena unguis-cati

Tecoma stans

Bixa orellana

17

BOMBACACEAE

Ceiba pentandra (L.) Gaertn.*

Local names: Silk Cotton Tree, 'Fromager' (Dominica, Fr. West Indies).

Native to the American tropics (i) it is found in drier coastal areas.

This tree grows up to 40 m and is deciduous. It is widely branched and sometimes buttressed and may have prickles on the trunk or branches. Its leaves are palmate. The flowers are pale ivory with silky petals. Its pods are about 10–30 cm long and open to reveal brown silky fibre of very soft texture, attached to tiny round seeds. Often, one's first introduction to the tree is the evidence of piles of glossy fibres along paths or areas where the tree grows.

Although soft, this fibre is not much used because it is difficult to obtain in large quantities and is difficult to clean. In Dominica the tree is said to be the home of spirits. Caribs rarely used the floss because they believed their sleep would be haunted (xix). However, the legend says that the wood may be cut between March and May when the spirits are absent.

Medicinal uses It was said to be used in refreshing baths to relieve fatigue, and to counteract poisoning. It was also used as a tea against colic and inflammation, taken as infusions.

Ochroma pyramidale (Cav.) Urb.*

Local names: Balsa, 'Bois Flot' (Dominica, Fr. West Indies, St. Lucia).

Native to the West Indies (i). Found in forests and secondary growth at middle to upper elevations in Dominica.

A fairly quick-growing tree. It may grow to 20 m. The flowers are large, solitary, and distinctive, with pale yellow petals and dark maroon, slightly hairy sepals.

The fibres attached to the seeds are used in stuffing pillows and cushions. *Ochroma pyramidale* fibre is used more frequently than that of *Ceiba pentandra*, although its floss is not as fine. The wood is used primarily for floats.

This family is closely related to the Malvaceae or Hibiscus family.

BORAGINACEAE

Cordia collococca L.

Local names: Manjack (St. Thomas); Wild Clammy Cherry (Barbados).

Native to the West Indies, from Mexico to South America (i), found in coastal elevations.

A medium-sized tree (up to 10 m) which has ovate leaves up to 18 cm long, and a spreading habit. The flowers are in clusters and are small and white. The berries are bright scarlet, round and shiny, with a fleshy, sticky pulp.

(*Cordia sebestena* is widely used as a cultivated ornamental for its red flowers, growing well in dry coastal conditions as seen in Antigua, St. Maarten and St. Thomas, where it has been grown as a shade tree).

Medicinal uses The leaves were used in a tea as a soporific. The mucilage from the berries was used for diarrhoea and dysentery.

Ceiba pentandra

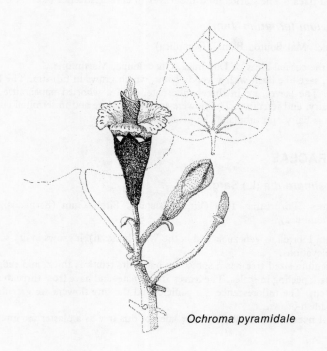

Ochroma pyramidale

BORAGINACEAE (continued)

Cordia curassavica (Jacq.) R. & S.*

Local names: Black Sage (Barbados); 'Verveine Bord la Mer', 'Absinthe Bord la Mer' (Martinique and Guadeloupe).

Found in open wastes at lower elevations in St. Vincent, Barbados.

A woody shrub about 1–2 m high with a spreading habit. The leaves are a soft grey-green and are aromatic, giving an ornamental appearance. The flowers are small, white, and borne in terminal spikes. The fruit is white at first, then turns red.

Cordia laevigata Lam.

Local name: 'Coco Poule' (Dominica).

Found in the Greater Antilles, Virgin Islands, Central America (i). This tree grows at lower to middle elevations, usually in windbreaks and in secondary forests.

A medium-sized tree with a greyish trunk and straggling habit. The ovate leaves are large (from 18–30 cm long), rough textured, and closely alternate on the stem. Usually there is a smaller leaf just below, which gives the appearance of groups of three leaves together. The margins are entire. The old leaves turn red before dropping off. The small white flowers are in panicles with five stamens and five petals, which attract honeybees, when first opened.

Medicinal uses The Caribs used the leaves to cure headaches (xix).

Heliotropium ternatum Vahl.*

Local name: 'Mal Bouton Blanc' (Dominica).

Found in the coastal areas of Dominica, Guadeloupe, Martinique.

A small, sage-like bush with a woody habit, which grows in full sun. The bark is grey and scaly. The leaves are in threes, alternate, with a whorled appearance. These are roughly hairy, and retentive. The flowers are small, white and in terminal curled spikes. The calyx is hairy and tubular. The petals are lobed.

BURSERACEAE

Bursera simarouba (L.) Sarg.

Local names: Turpentine Tree (Virgin Islands); Birch Gum (Barbados); 'Gommier Rouge' (Dominica).

Found from Florida to Venezuela and in the West Indies (i). It grows in dry scrubby areas at lower elevations.

This medium-sized tree has a spreading habit. Its trunk is thick, and reddish-brown, with the bark peeling in scales. The leaves are pinnate, and have from three to nine leaflets on each stem. The inflorescence is a panicle, and the tiny flowers are greenish-red. The fruit is reddish-brown.

Medicinal uses The Caribs used the latex of this tree as a plaster for internal bruises (xix).

Cordia curassavica

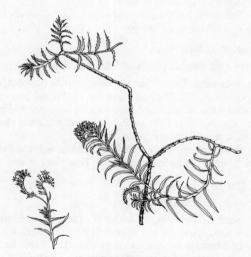

Heliotropium ternatum

BURSERACEAE (continued)

Dacryodes excelsa Gr.

Local name: 'Gommier' (Dominica, St. Lucia, Fr. West Indies).

Found in some of the rain forests of the Lesser Antilles.

It grows up to 40 m high and is easily distinguished by its straight, cylindrical, massive trunk. The bark is greyish. The leaves are unequally pinnate, usually five to seven on a stalk. The leaflets are dark green and leathery in texture, with an elliptic shape. The greenish flowers are very small, and have three or sometimes four small petals. Male and female flowers occur on separate trees. The fruit is a drupe with a green or purple skin, very much resembling an olive in shape and appearance. These usually fall to the ground in clusters of three or more. Because of its height, it is difficult to identify the flowers, which bloom in July, so identification is best made by examination of the trunk, or the fallen fruit.

'Gommier' is noted for its resin which flows from the bark when cut. This thick substance has a strong incense-like scent and is excellent for lighting charcoal and as a 'flambeau' or torchlight.

Medicinal uses   The Caribs used the gum for toothache and also to relieve shortness of breath (xix).

CACTACEAE

Opuntia cochenillifera (L.) Mill.*

Local names: Prickly Pear; 'Rachette'.

Found in the same areas as *O. dillenii*, this species is slightly larger, the segments are more fleshy and the spines blunter, making it easier to handle.

Medicinal uses   This plant has the same uses as *O. dillenii*. It was also used as a tea for 'inflammation' and as a bath and a shampoo.

Opuntia dillenii (Ker.-Gawl) Haw.*

Local names: Prickly Pear (Barbados); 'Mal Rachette' (Dominica).

Found widely from south-east United States to northern S. America, in the West Indies, Mexico, the Canary Islands, S. India and Australia (i). Found near the coasts.

The fleshy segments are quite thin, and covered with long sharp spines regularly spaced over each segment. The flower is light yellow, sometimes pink at the base.

It is sometimes used as a boundary marker.

Medicinal uses   The segments of the stem have been boiled into a tea for inflammation or as a bath to which other plants may be added. This plant is also used to wash hair.

Opuntia tuna (L.) Mill.

Local names: Prickly Pear; Cochineel.

This plant is also found in coastal regions, and is commonly used as an ornamental.

It is shrubby, also with spines. It has yellow flowers and dark red fruit.

Medicinal uses   In Barbados (iv) this had wide uses. The stems were peeled and boiled in sea water for gripes, ulcers, and to stop menstruation. The juice of the fruit was used as a colouring agent. The leaves were cut and applied to wounds. The pulp gives a soapy lather which was used for hair washing, as in the other species. The leaves and stems, baked and sliced, were used for swellings or headaches. For soreness of constipation and pains they could be used with salt as a poultice (vii).

22

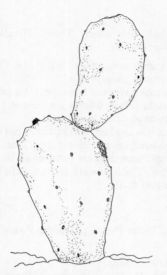

Opuntia cochenillifera

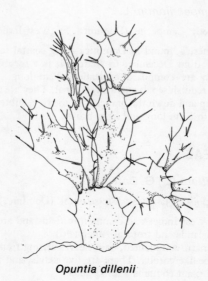

Opuntia dillenii

CAESALPINIACEAE

Cassia alata L.*

Local names: Impetigo Bush (Barbados); 'Desay' (Dominica); Christmas Candles (Barbados); 'Dartier' (Fr. West Indies).

Native to American tropics and now naturalized in the Old World. Found at most elevations but does best in coastal zones.

This shrub bears large pinnate leaves and has a spreading habit. The texture of the stem and leaves is coarse. The bright yellow flowers are erect and candle-like.

This shrub is used mainly as an ornamental.

Medicinal uses The plant was used for skin maladies (xi) and scurf. When steeped in water the leaves were considered an excellent gargle. The pulp of the seed pod was used as a purgative. The seeds themselves are **poisonous,** so the pulp should be scraped off by hand and not boiled off. The leaves are used in baths to clean skin, and in Marie Galante in teas for hypertension (xvi).

Cassia occidentalis L.*

Local names: 'Café Puant; Z'herbe Pienne'; 'Z'herbe Piante'; 'Café Mocha' (Dominica, Fr. West Indies).

Found at coastal and middle elevations.

A spindly, herbaceous plant. The leaves are pinnate, alternate on the stem, and give off a strong odour when crushed. The flowers are in a small cluster, each with five curled yellow petals. The thin pods are many-segmented, and turn brown when mature.

Medicinal uses The seeds were often roasted as a substitute for coffee, and teas were made for colds and stomach disorders from the flowers and roots (xix). When soaked in warm water the roots were used for skin disorders and for swelling of the legs.

Haematoxylon campechianum L.

Local names: Logwood; 'Campêche' (Dominica, Fr. West Indies).

Native of Central America, found in Dominica along coastal roads.

This plant blooms from December to March. It is a scrubby small tree with small pinnate leaves which are composed of leaflets 2 cm long. The yellow flowers are numerous with small reddish sepals and a sweet scent. They are tiny and occur in racemes of about 5 cm long, up and down the branches of the tree, subtended by a leaf. The pods are thin and flat, up to 5 cm long and light brown.

CAMPANULACEAE

Hippobroma longiflora (L.) G. Don

Local names: Horse poison (Jamaica); 'Pipe Zombi' (Dominica).

A herb found in the West Indies and common in fields and areas of cultivation. Found also in some Pacific islands and tropical America (i).

A small herb with narrow, indented, hairy leaves arising from the ground. The single flower has a long tube-like corolla. There are five petals, and the flower is white. The calyx is toothed. The plant contains a milky juice.

Medicinal uses The leaves were used by the Caribs as a poultice for foot injuries (xix).

Cassia alata

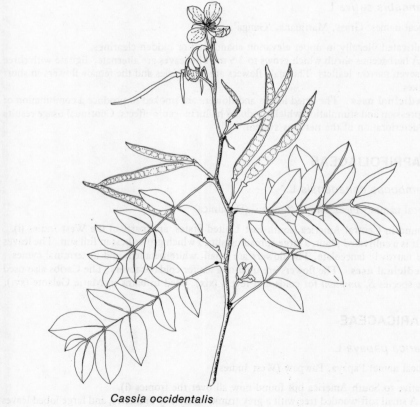

Cassia occidentalis

25

CAMPANULACEAE (continued)

Lobelia flavescens (Presl.) E. Wimm*

Local name: 'Fleur Montagne' (Martinique and Guadeloupe).

Found at upper elevations, on summits of mountains.
This is a low shrubby plant with whorled lanceolate leaves and toothed margins. The white flowers are on spikes, and the petals are divided into five lobes. The stamens are purple-tipped.

Lobelia persicaefolia Lam.*

Found at middle elevations in poor soil conditions, usually growing in full sun.
This shrub is low and straggling. The leaves are lanceolate, and toothed at the margin. The flowers are brilliant scarlet, divided into five lobes and solitary on long stems. There are two bracts far below the stem.

CANNABACEAE

Cannabis sativa L.

Local names: Grass, Marijuana, 'Ganga'.

Cultivated illegally in upper elevation abandoned or hidden clearings.
A herbaceous shrub which grows to 3.5 m. The leaves are alternate, digitate with three to seven narrow leaflets. The male flowers are in panicles and the female flowers in short spikes.
Medicinal uses   The dried leaves and flowers are smoked to produce a combination of depression and stimulation which produces hallucinogenic effects. Continual usage results in deterioration of the nervous system.

CAPRIFOLIACEAE

Sambucus canadensis L.*

Local names: 'Fleur Sirop'; 'Suyeau' (Dominica).

Found in Central America, south-east United States, and parts of the West Indies (i).
It is a cultivated plant, a large spreading shrub which grows best in full sun. The leaves are narrowly lanceolate. The flowers are small, white-petalled, and in terminal cymes.
Medicinal uses   The flowers are used in a tea for colds or fevers. The Caribs also used the species *S. simpsoni* for colds and fevers (xix). Used in baths in Marie Galante (xvi).

CARICACEAE

Carica papaya L.

Local name: Papaya, Pawpaw (West Indies).

Native to South America but found now all over the tropics (i).
A small soft-wooded tree with a grey trunk, rarely any branches, and large lobed leaves growing in a whorl from the top. These are about 30–60 cm in diameter. The pale yellow male and female flowers are on separate trees — the males on long racemes and the females

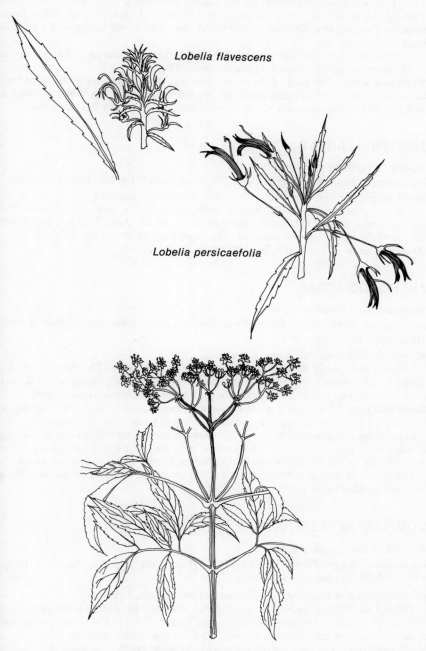

Lobelia flavescens

Lobelia persicaefolia

Sambucus canadensis

27

close to the trunk near the leaves. All parts of the tree have a milky latex. The fruit is spherical and differs in size according to the variety, but all turn deep yellow upon maturity. They are best if picked slightly before ripe, then scored lightly with a knife to allow latex to seep out.

Commercial meat tenderizers are obtained from the latex, but one can wrap meat in one of the leaves for several hours to obtain the same effect.

Medicinal uses The juice of the leaves, stem and roots is used for boils, ringworm, warts and worms (iv). The seeds, eaten raw, are said to cause abortion. The green fruit is used for colds and dyspepsia, and the leaves can be applied locally for ulcers (viii).

CAROPHYLLACEAE

Drymaria cordata (L.) Willd.*

Local names: 'Moulon', 'Mouron blanc' (Fr. West Indies); Timéron (Marie Galante).

Found generally in the tropics and commonly in lawns and pastures.

A small, creeping weed. The stalk is sticky and so is the small, round seed. The opposite leaves are very light green. These are small, round, about 5–10 mm broad, and like the whole plant, very soft textured. The flowers are white and insignificant.

Medicinal uses Used in teas for colds (xvi).

CHENOPODIACEAE

Chenopodium ambrosoides L.*

Local names: Worm Weed, Bitter Weed, 'Simen Contra' (Dominica); 'Herbe à Vers' (Guadeloupe and Martinique).

Widespread distribution in warm countries (i).

An erect herb about 90 cm high, with very odourous leaves. The older leaves are lanceolate and toothed, the younger are entire margined. The greenish-white flowers grow in clusters on spikes. There are male and female flowers on the same plant.

Medicinal uses The plant is used basically as a vermifuge all through the West Indies. The leaves are used in cooling teas and in Carib ritual baths, after childbirth (xix).

In Barbados it was steeped in rum which absorbed more of the oils and this produced a stronger vermifuge (viii). It was also used in milk for children. If used to its fullest potential, the results are gastro-intestinal disorders, irritant effects, coma and **death**.

The plant was used as a vermifuge only at the descent of the moon in Marie Galante (xvi), and in Martinique (x).

COMPOSITAE (ASTERACEAE)

Ageratum conyzoides L.*

Local names: 'Bouton Blanc', 'Petit Pain Doux' (Dominica); 'Herbe aux Sorciers', 'Herbe à Femme' (Martinique and Guadeloupe).

Found generally in warm countries in pastures and neglected gardens.

This is a small straggling herb with tufted blue or white (rare) flowers borne in terminal clusters. They have no ray flowers or discernible 'petals'. The leaves are ovate and toothed at the margins, and are velvety in texture. They are opposite on the stem.

Medicinal uses Used as a constituent in a tea for colds. Formerly it was widely used in Guadeloupe and Martinique (xi) as a diuretic, for coughs and colds, and as a bath for skin eruptions. It was also used as a cooling tea.

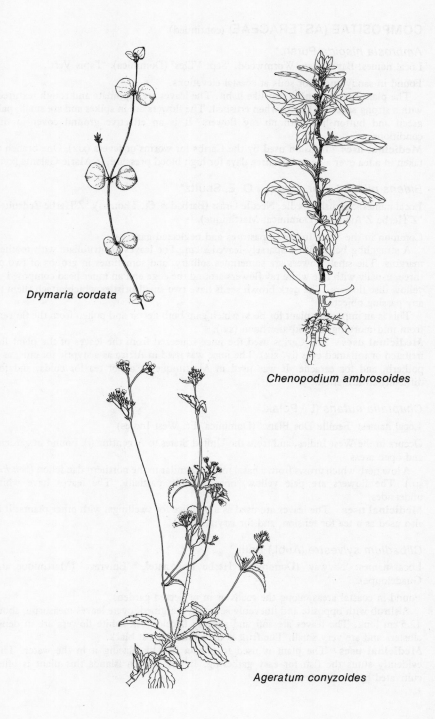

Drymaria cordata

Chenopodium ambrosoides

Ageratum conyzoides

COMPOSITAE (ASTERACEAE) (continued)

Ambrosia hispida Pursh.*

Local names: Bay Tansy, Wormwood, 'Sept Villes' (Dominica); 'Tapis Vert'.

Found in sandy and poor soils at coastal elevations.

The plant has a creeping mat-like habit. The leaves are bipinnate and rough textured, with a strong sage-like aroma when crushed. The flowers are in spikes and are small, pale green and button-like with no ray flowers. It is an effective ground cover in dry conditions.

Medicinal uses Infusion used by the Caribs for worms or fevers (xix). One branch is taken in a tea over a period of three days for high blood pressure, in Marie Galante (xvi).

Bidens pilosa L. var Alba (L.) O. E. Shultz*

Local names: Spanish Needle, Needle Grass (Barbados, St. Thomas); 'Z'Herbe Zedruite', 'Z'Herbe Z'Aiguille' (Dominica, Martinique).

Common in the West Indies in pastures and neglected gardens.

A straggling herb with a sparsely-leaved stem. The leaves are trifoliate with toothed margins. The small flowers are sometimes solitary, and sometimes in groups of two or three, usually with five white ray flowers around the edge and an inner head composed of yellow disc flowers. The dark brown seeds have two small bristles which attach them to any passing object.

This is an important plant for bees, which gain both nectar and pollen from the flowers from mid-morning, to mid-afternoon (xxi).

Medicinal uses The Caribs used the juice squeezed from the leaves of the plant for irritated or inflamed eyes (iv, xix). The juice was used in Africa as a styptic for cuts, as a potherb, and for earache. It was used in Martinique as a hot tea for colds, and for difficulty in urination.

Chaptalia nutans (L.) Polak.*

Local names: 'Feuille Dos Blanc' (Dominica, Fr. West Indies).

Occurs in the West Indies, and from the United States to Argentina (i). Found in gardens and open areas.

A low herb which grows from a basal rosette, similar to the northern dandelion (*Sonchus* sp.). The flowers are pale yellow, opening only partially. The leaves have white undersides.

Medicinal uses The leaves are used as a poultice for swellings, with other plants. It is also used as a tea for tension, and for laryngitis (x).

Clibadium sylvestre (Aubl.) Baill.*

Local names: 'Nivway' (Dominica); 'Herbe Enivrante', 'Enivrage' (Martinique and Guadeloupe).

Found in coastal areas, along the roads or in neglected gardens.

A shrub with opposite and unevenly serrated, margined, ovate leaves measuring about 12.5 cm long. The leaves are soft and slightly flaccid. The white flowers are in dense clusters and are very small. The fruit is dark blue, almost black.

Medicinal uses The plant is used to poison fish by placing it in the water. This evidently stuns the fish for easy gathering. In the French islands this plant is often cultivated by the fishermen.

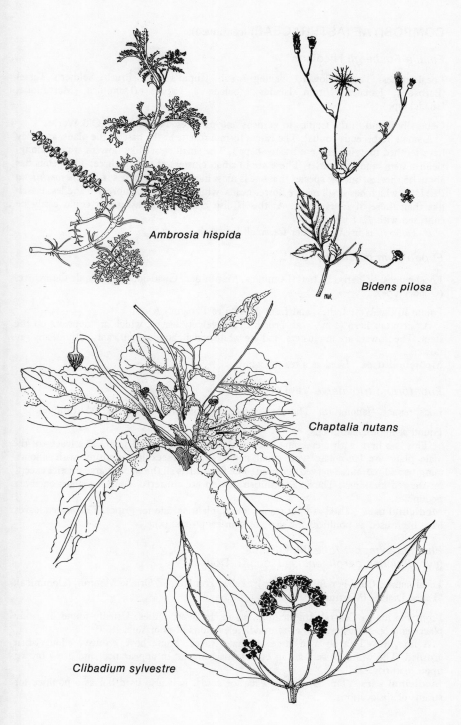

Ambrosia hispida

Bidens pilosa

Chaptalia nutans

Clibadium sylvestre

31

COMPOSITAE (ASTERACEAE) (continued)

Emelia Fosbergii Nicolson*

Local names: Tassel Flower, Shaving Brush, Cupid's Paint Brush, Soldier's Tassel (Barbados, Jamaica, Virgin Islands); 'Salade à Lapins' (Dominica, Martinique, Guadeloupe).

Generally found in the tropics in gardens and pastures. Native to the Old World.

A small herbaceous annual. The leaves form a rosette at the base of the plant, and they have toothed margins and are lyrate-shaped. The small tassel-like flowers are on a long, slender stem, and are scarlet. These are in small clusters of two or three. The genus has about 45 species, but the species that one is most likely to come across is *Emelia sonchifolia* (L.) DC., which has small mauve flower heads, which are more compressed and less tassel-like than those of *E. Fosbergii*. As this is also common in pastures it could easily be confused with *E. Fosbergii*.

E. Fosbergii is used as fodder for small livestock.

Eupatorium macrophyllum L.*

Local names: 'Z'herbe à Chat' (Dominica, Martinique, Guadeloupe); 'Grande Guimauve' (Guadeloupe).

Found in the West Indies, and from Mexico to Paraguay (i).

A fairly tall herb (1 m), with broad, ovate, velvety leaves, which are opposite on the stem. The flowers are in clusters, and are white, small, and insignificant, without any ray flowers.

Medicinal uses Used as a tea for wind.

Eupatorium triplinerve Vahl.*

Local names: 'Japona' (St. Thomas); 'Japanne' (Dominica); 'Diapana' (Martinique).

Found at middle elevations.

This is a herb with a creeping habit and a maroon coloured stem. The leaves of the male plant are lanceolate with small petioles, almost sessile, opposite, with smooth margins and red, three-nerved venation. The female has all the same characteristics except for the red markings. The small, whitish flowers are clustered along the stems on short peduncles.

Medicinal uses Used as a hot tea for fevers, pleurisy, and for gripes. The heated leaves have been used as poultices for sores or skin eruptions. (x).

Pluchea symphytifolia (Mill.) Gillis*
(syn. Pluchea carolinensis (Jacq.) G. Don)

Local names: Cure For All (Barbados); 'Tabac Zombie', 'Z'Orielle Mouton' (Dominica); 'Tabac Diable' (Martinique).

From Florida to Central America (i), and in the West Indies. Usually found in sandy places and open wastes. Grows at middle elevations in poor soils.

A furry-leaved shrub with large alternate leaves. These have a musty odour when crushed. The white flowers are in cymes, insignificant in appearance, and grow brown upon maturity.

Medicinal uses This is used in a tea for colds. It is also used hot as a poultice for strains or dislocations.

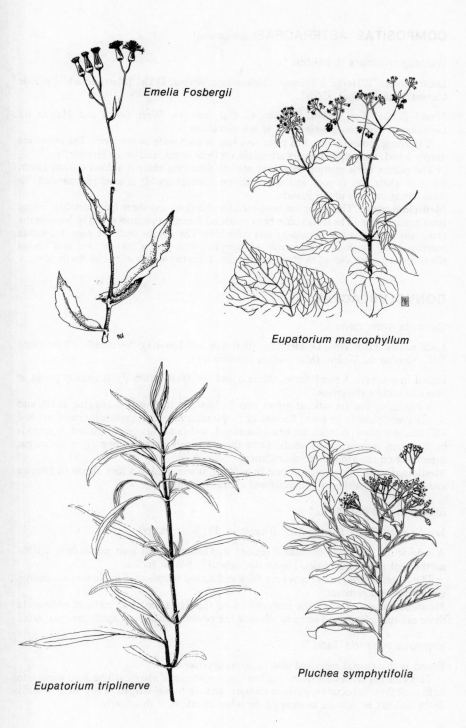

Emelia Fosbergii

Eupatorium macrophyllum

Eupatorium triplinerve

Pluchea symphytifolia

33

COMPOSITAE (ASTERACEAE) (continued)

Wedelia trilobata (L.) Hitch.*

Local names: 'Z'Herbe à Femme' (Dominica); 'Bouton D'Or' (Martinique); 'Patte de Canard' (Guadeloupe).

Found in the West Indies, Florida, Central America, West Africa and Hawaii (i). Common in pastures, growing best at low elevations.

A creeping ground cover, this plant will root at each node on the stem. The leaves are deeply lobed, the yellow flowers are single on their stems, and occur terminally.

The plant's thick matting habit and profuse flowering make it a good ground cover, effective where soil is poor and soil retention is necessary. It is used commercially in landscaping in some of the islands.

Medicinal uses This plant has been used for abortions and sheep have lost their young from grazing on it. The leaves have been pounded for use as poultices, and in Jamaica the plant was used as a tea for coughs and colds (iv). The Caribs used the pounded leaves together with those of 'Immortelle' (*Erythrina corallodenron* var. *bicolor*) and Quina (*Exostemma sanctae-luciae*) to clear the placenta from the womb after childbirth (xix).

CONVOLVULACEAE

Cuscuta americana L.

Local names: Love Vine, Dodder Vine (Barbados and Jamaica); 'Vermicelle', 'Liane sans Fin', 'Cordon du Violon' (Martinique, Guadeloupe).

Found in southern United States, Mexico and the West Indies (i). It usually grows at lower to middle elevations.

A straggling leafless vine, it grows over hedges and trees with a clinging habit, and covers every plant in its path. Because of its parasitic nature, it eventually smothers and kills the host plant. It looks like orange seaweed, and can easily be recognized. It spreads by the stem rooting, or by seeds. Once the plant gets onto the host, it is no longer dependent on the ground for nourishment.

Medicinal uses A pernicious pest in gardens, it was used as a love charm in Jamaica and was used in Africa as a diuretic and a laxative. (viii).

Ipomoea pes-caprae L.*

Local name: 'Patate Bord la Mer' (Dominica, Fr. West Indies).

A vine found on sandy or coastal regions in Dominica, and seen particularly on the northern or windward coasts. Found throughout the West Indies.

The leaves of this ground cover are thick and almost succulent. The flowers are mauve, and fairly wind resistant.

Medicinal uses The Caribs frequently used the leaves for ritual baths in addition to those of other plants, and often to alleviate the power of malignant spells, or 'piai' (xix).

Ipomoea repanda Jacq.*

Found in dry coastal areas and also in secondary forests.

This vine has somewhat fleshy cordate leaves with entire margins. The deep maroon to light pink flowers occur on stalks in clusters, and are funnel-shaped. They are also fairly fleshy and not as delicate as some of the other members of this family.

Wedelia trilobata

Ipomoea pes-caprae

Ipomoea repanda

CONVOLVULACEAE (continued)

Ipomoea hederifolia L.*

A native of tropical America (i), this vine has been found growing in coastal areas in St. Thomas, and is now an escape in gardens in Dominica. It grows best in neglected areas in full sun.

An annual vine, it has cordate leaves, the more mature ones occasionally lobed. The funnel-shaped flowers occur in clusters on short stems, and are a brilliant scarlet. The four small, dark brown seeds are in oval capsules.

It is a good ornamental for a small lattice or screen. Another species with similar flowers is *Ipomoea quamoclit* which has very feathery pinnate leaves and is often cultivated.

Ipomoea tiliacea (Willd.) Choisy*

Local names: 'Caapi', 'Caapi Doux' (Dominica); 'Patate Bâtard', 'Patate Sauvage' (Fr. West Indies).

Found from Florida to Peru and in the West Indies in neglected areas, commonly covering trees and shrubs.

This vine has simple cordate leaves and tendrils. The flowers are light mauve and at a distance seem almost white. The centres of the corollas are purple, and the berries are in clusters of four. It is readily identified by its habit of covering orchard trees with a thick mat.

The vine is used as a fodder for livestock.

Medicinal uses   The Caribs used the leaves in baths for good luck (xix).

Porana paniculata Roxb.*

Local names: White Coralilla (Barbados); Christmas Vine (Jamaica); 'Muguet' (Fr. West Indies).

Originally from India and Malaysia, it was introduced from Brazil and Uruguay to Martinique in 1875 (xi). It blooms at lower elevations in December.

It has alternate, soft, cordate leaves with entire margins. The flowers occur in racemes and are white with a sweet scent.

This vine was evidently once cultivated, but it is now mostly seen as an escape from gardens.

CRASSULACEAE

Kalanchoe pinnata (Lam.) Pers.*
(syn. Bryophyllum pinnatum (Lam.) Oken.)

Local names: Leaf of Life; 'Herbe Mal Tête' (Dominica).

Native of Madagascar and now widespread in the tropics (i). Found in gardens, roadsides and pastures at lower to middle elevations.

An erect herb with very fleshy stems, it grows up to 1 m high. The stems are slightly reddish, and the leaves, also fleshy, have crenate margins. Small plants readily grow along these margins. The flowers are in clusters with light-green calyxes and red corollas, which droop.

This plant is ornamental in rockeries in dry areas and spreads readily.

Medicinal uses   The leaves were used in a cooling tea and also as a poultice. This was prepared by breaking the leaf and applying the liquid on the head or on a sore.

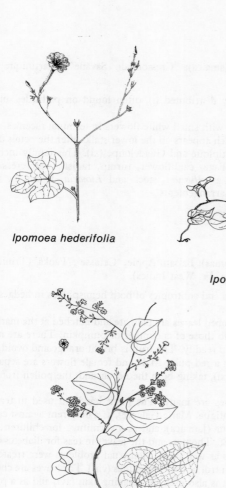

Ipomoea hederifolia

Ipomoea tiliacea

Porana paniculata

Kalanchoe pinnata

CRUCIFERAE

Lepidium virginicum L.*

Local names: Wild Peppergrass (Jamaica); 'Cresson de Savane' (Martinique and Guadeloupe).

Native of North America and widely distributed (i), often found on pathsides and in neglected gardens.

This annual herb is finely-textured with small white flowers in terminal racemes, each bearing a flat seed upon maturity, which appears on the lower stalks after the petals drop.

It was often used in salads in Martinique and Guadeloupe (xi). This family includes many well-known vegetables — cabbage, cauliflower, turnip, radish, watercress and broccoli and also ornamentals such as *Alyssum*, stock and *Lunaria*. It can be used ornamentally, after drying, in flower arrangements.

CUCURBITACEAE

Mormordica charantia L.*

Local names: Maiden Apple (St. Thomas); Balsam Apple, 'Cerasse', 'Paoka' (Trinidad); 'Pomme Coolee', 'Cocouli' (Dominica, Fr. West Indies).

Native to Asia (i), found in the tropics and subtropics of both hemispheres in hedges and pathsides, usually at lower elevations.

An annual herbaceous plant. The lobed leaves are palmate and notched at the margins. The flowers are pale yellow, similar to those of cucumber or pumpkin. There are many tendrils which enable the vine to creep readily. The fruit is bright orange and ovoid, and the black seeds inside are coated with a red pulp. Male and female flowers are separate.

Bees are attracted to the flowers (xxi), taking both the nectar and, the pollen from the male flowers.

Medicinal uses   The medicinal uses are many. The plant has been used in teas for colds, fevers, high blood pressure (Antigua, Marie Galante), as a deterrent against cancer by checking the growth of the tumour (Jamaica), and as a vermifuge for children. The fruit was used as an abortifacient in St. Thomas, and the leaves in teas for diabetes (xiv). The leaves may be used for poultices as well (xix). Menstrual problems were treated by this plant, and it was used for birth control in the Grenadines (viii). The leaves are chewed for sore throats, and used for a tea. It is also used for cleaning skin (xvi) and as a purge, taken for several days consecutively to wash impurities from the system.

The young fruit was used in curries in the East.

EUPHORBIACEAE

Croton flavens L.

Local names: 'Baume', 'Petit Baume', 'Coupai' (Fr. West Indies).

A bush or small tree found growing along the roadsides in Dominica near Roget, Castle. Comfort and going toward Soufriere. It grows in dry conditions at low elevations.

It has a smooth grey bark and the younger leaves are yellowish. The distinguishing features are occasional very bright orange leaves on the bush. These leaves are ovate and about 7 cm long. The small and inconspicuous flowers are white.

The wood is often used for constructing fences.

Medicinal uses   The aromatic sap has been distilled for use in toiletries. It has also been used on sores (xi).

Lepidium virginicum

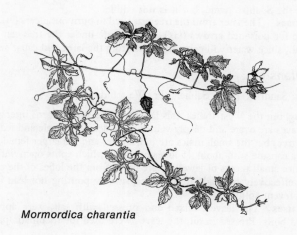

Mormordica charantia

EUPHORBIACEAE (continued)

Euphorbia hirta (L.) Millsp.*

Local names: 'Mal Nommée Vrai', 'Z'Herbe Mal Nommée' (Guadeloupe).

This small spreading weed with prostrate branches is found in neglected areas, gardens and grasslands in the tropics and subtropics (i).

The leaves and the stem are reddish, and the flowers are greenish-red or mauve.

Medicinal uses   It was widely used in Guadeloupe (xi) as a tea for fevers, to induce regular urination (xi), and later, for asthma and bronchitis.

Euphorbia prostrata Ait.*

Local names: 'Paille Terre', 'Petite Teigne Noire' (Guadeloupe).

Found in coastal areas in rubble, on poor soil, frequently in stairways and old walls.

This is a very small prostrate herb with tiny opposite leaves and reddish stems.

Medicinal uses   This is used in Dominica widely to combat 'flu and in Guadeloupe as a 'tisane' for nursing mothers to purify the blood and improve their milk. It was also used against dysentery (xi).

Hippomane mancinella L.

Local names: 'Manchineel' (West Indies); 'Madjini' (Dominica); 'Mancenillier' (Fr. West Indies).

Found in the West Indies, West Africa, on the Pacific coast of Central America and from Florida to Venezuela (i).

A tree which grows up to 10 m high with a broad and spreading crown. The leaves are bright green, alternate, and slightly toothed, about 7.5–10 cm long. The flowers are greenish, small and insignificant. The fruit is round, shiny, green and it looks like a small apple, hence the Spanish name, but it is **not** edible.

Medicinal uses   The latex from the tree was used to burn out ulcers (vi), and the Caribs used this sap for poisoned arrows (xix). Even sitting under the tree can be **dangerous** after a rainfall, since water will carry the toxin from the latex and cause severe blistering.

Hura crepitans L.*

Local names: Sandbox Tree, 'Sablier' (Fr. West Indies).

Found throughout the West Indies, this large tree has small thorns interspersed over the bark. The leaves are ovate and deeply veined. The deep red, male and female flowers are on the same tree, but the small males are in a cluster and the larger females, are separate. The fruit is a capsule with about 15 lobes or more, which splits open with an explosion.

In Dominica small pieces of jewellery are made from the lobes of the capsules. Paperweights can effectively be made from the capsule by pouring hot lead into the central opening to prevent splitting.

Medicinal uses   The leaves, mixed and pressed with salt, were applied locally on swellings and boils. Pressed in oil, the leaves were used for rheumatic pain (xi).

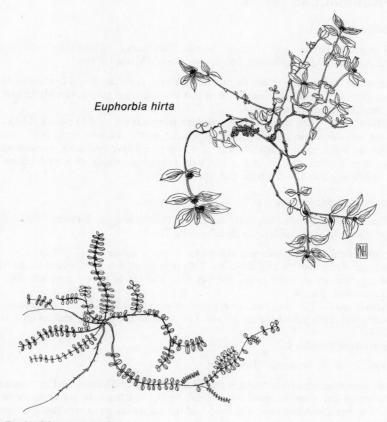

Euphorbia hirta

Euphorbia prostrata

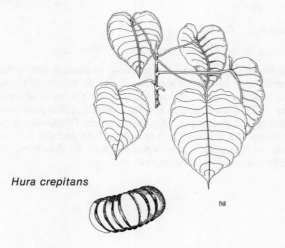

Hura crepitans

EUPHORBIACEAE (continued)

Jatropha curcas L.*

Local names: Physic Nut (Virgin Islands, Barbados); 'Médecinier' (Dominica); 'Médecinier Blanc' (Guadeloupe); Médecinier Barrié (Marie Galante).

A large shrub with heart-shaped, slightly lobed leaves, generally found in the tropics and subtropics in dry areas. The fruit is green and smooth when young, turning yellow and then deep brown at maturity.

Medicinal uses   The seeds contain a **violent purgative** if used to excess. The Caribs used the seeds as a purgative or emetic depending on the quantity given (xix). In other cultures the leaves were boiled for general sickness, and specifically for stomach ulcers. These were bruised, heated and made into a poultice which alleviated ulcers and rheumatism (iv), and also for boils and leprosy (viii).

Jatropha gossypifolia L.*

Local names: Wild Physic Nut, Belly Ache Bush (Virgin Islands, Barbados); 'Médecinier Rouge' (Guadeloupe); 'Médecinier Bené' (Marie Galante).

Generally found in the tropics and subtropics in sandy areas.
   A smaller shrub than the last, about 60–120 cm high, with deeply lobed alternate leaves and many hairs all over the stems and petioles. The leaves have a purplish cast. The flowers are deep purple.

Medicinal uses   The same medicinal purposes as applied to *J. curcas*. In Barbados (vii) the leaves and seeds were used for gripes. The oil from the seeds resembles olive oil.

Margaritaria nobilis L.*

Local name: 'Mille Branches' (Dominica).

A small to medium-sized tree found at middle elevations and distinguished by branches of elliptic leaves like sheaths. Small whitish dots, which are lenticels, are apparent on the stems. The tree illustrated here is in fruit, and the numerous green capsules split open to show four navy blue segments within. The flowers are green and extremely small, in little clusters along the branches.

Phyllanthus tenellus Roxb.*

Local names: 'Grain en bas Feuille' (Dominica, Fr. West Indies).

Native to the Mascarene Islands (i) it is now found in gardens and orchards.
   A single-stemmed small herb. Its distinctive feature is the habit of the seeds, which are round and green or slightly red. They are borne under the stalks of the pinnate leaves.

Medicinal uses   The Caribs used it as a tea to produce abortions (xix). The plant is **poisonous** to rabbits and guinea pigs. A similar species, *Phyllanthus niruri*, was used in Guadeloupe for fevers (xi). It is used as a refreshing bath and also with other plants in a bath against 'piai' or 'l'amarage'. It is reputedly good for diarrhoea, the female plant (smaller and red) being considered better than the male.

Jatropha curcas

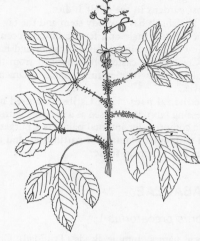

Jatropha gossypifolia

Margaritaria nobilis

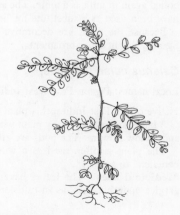

Phyllanthus tenellus

43

EUPHORBIACEAE (continued)

Ricinus communis L.*

Local names: Castor Oil, 'Huile Cawapat', 'Cawapate' (Dominica and Martinique); 'Huile de Ricin' (Martinique).

A large shrub, native of the Old World tropics and now widely distributed, it is found in most gardens in the West Indies.

The shrub has a woody stem and the big palmate leaves are sometimes reddish-tinged. The whitish male and female flowers occur on erect terminal and axillary spikes. The shiny seeds are in burr-like capsules and have distinctive woodgrain markings.

Commercial castor oil is derived from these seeds. It has the advantage of not congealing at low temperatures. Some ornamental varieties have deep maroon seeds and red leaves.

Medicinal uses The Caribs have used it as a constituent for body paint and for hair dressing (xix). It was used as a massage with other plants, the leaves being heated and used as a compress for internal pain. As a purgative it is well known, but it was also used as a tea with other plants after parturition, and also for gonorrhoea. The seeds contain abrin, which is extemely **toxic**.

FABACEAE

Abrus precatorius L.*

Local names: Jumbie Beads (Trinidad); Crab's Eyes, Wild Liquorice (Jamaica); 'Grain d'Eglise' (Dominica); 'Reglisse' (Fr. West Indies).

Found throughout the tropics (i), this woody climber likes dry soil and coastal conditions.

The leaves are pinnate, and the small mauve flowers are in small clusters along the stem. The brightly coloured red and black seeds are in hard twisted pods.

Medicinal uses The seeds are **poisonous**, but it is said that, if boiled, their toxic principle (toxalbumin) is destroyed. After this precautionary measure the seeds have been boiled again in milk as a tonic. The seeds are dangerous to livestock. The roots and stems have been used as a substitute for liquorice ('reglisse').

Because the seeds are decorative they have been used in jewellery, for the eyes of figures, and in other ornaments.

Cajanus cajan (L.) Millsp.*

Local names: Pigeon Pea (West Indies); 'Pois Angol' (Dominica, Fr. West Indies).

Native of the East Indies, this plant is commonly grown for its food value.

A large, spreading, bushy plant which bears silvery-green trifoliate leaves and small red and yellow flowers. The pods are green and soft-skinned.

The plant is often used as a windbreak in gardens, and its nitrogenous content is beneficial to soils.

Medicinal uses The leaves have been used in teas for colds (iv), and in Africa for gargles, mouthwashes, as an infusion for diarrhoea, and for smallpox (iv).

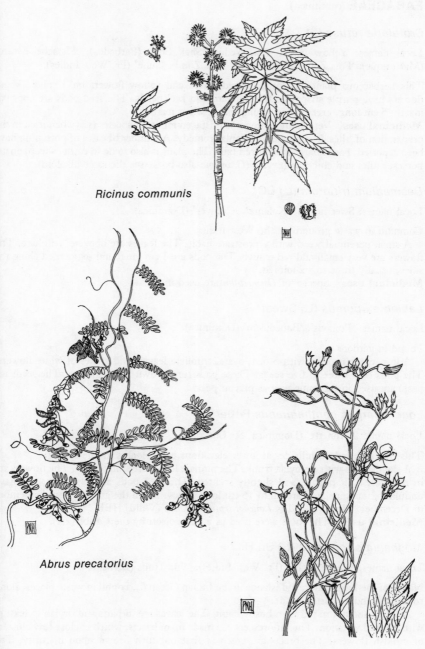

Ricinus communis

Abrus precatorius

Cajanus cajan

45

FABACEAE (continued)

Crotalaria retusa L.*

Local names: Yellow Sweet Pea, Yellow Shak Shak (Barbados); 'Pistache Bâtard' (Martinique); 'Pois Zombi', 'Sonnette', 'Pois Zombi Jaune' (Fr. West Indies).

This herbaceous plant has small, paired leaves and yellow flowers on a spike. These flowers have purple stripes on the standard and keel petals. The seed pods are approximately 5 cm long, green when young, black when mature.
Medicinal uses   Very **toxic**. Often it is mistaken for other species (xii) when used in the preparation of 'thés de sonnette', a popular remedy. A number of cases of poisoning have been reported, particularly of children (xii). The plant is also toxic to cattle, sheep, goats, horses, mules and chickens (xxvii). (*C. incana* also possesses the same alkaloid).

Desmodium triflorum (L.) DC*

Local names: Sweethearts, 'Kolante', 'Cacoyer' (Dominica).

Common in waste grounds in the West Indies.
   A small perennial weed with a prostrate habit. The leaves are obovate, trifoliate. The flowers are very small and red-mauve. The pods are 1 cm long, and segmented along the sides, usually three to six-jointed.
Medicinal uses   Species of *Desmodium* are used in baths.

Lablab purpureus (L.) Sweet*

Local names: 'Bonavis', 'Boucousou' (Dominica).

Found in gardens.
   A low, creeping vine with broad, ovate, trifoliate leaves, and spikes of white flowers. The pods bear three to five seeds. These pods turn brown when mature. The seeds (or peas) are used in the same way as pigeon peas.

Lonchocarpus benthamianus Pittier*

Local name: 'Savonette' (Dominica, St. Lucia).

This tree is found on hillsides at lower elevations, and blooms in late spring.
   A slender tree with a greyish trunk. The pinnate leaves are shiny green. The flowers are in racemes, light purple, and slightly scented. The seed pods are flat, wide, and have undulating margins, and are up to 15 cm long. These fall to the ground from September to December. Closely allied is *Lonchocarpus latifolius* (Willd) HBK.
Medicinal uses   The roots were used as a fish poison by the Caribs (xix).

Moghania strobilifera (L.) St. Hil.*

Local names: 'Herbe Sèche' (Fr. West Indies); Wild Hops (Barbados).

Native of the East Indies, and islands in the Indian Ocean (i). Found in waste places, along roadsides, and in partial shade.
   A shrub about 1 m tall with bushy habit. The leaves are in pairs and dryish in texture with marked venation. The inflorescence is made up of bracts, which enclose tiny clusters of pinkish flowers. The bracts are green at first but turn brown upon maturity. The flowers are replaced by tiny pods.
   Used with good effect in dried flower arrangements.

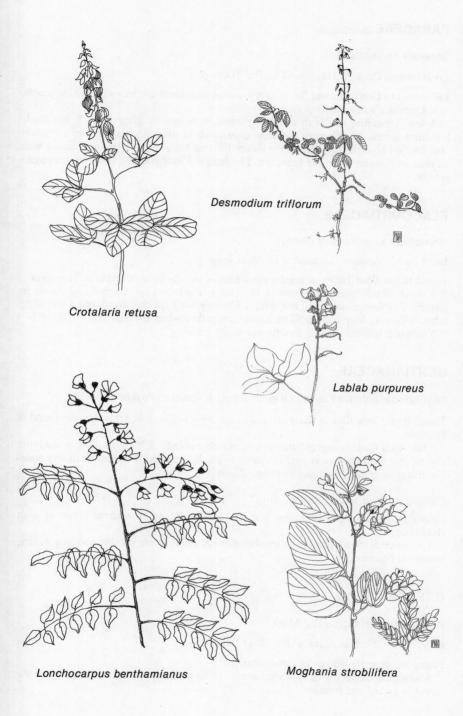

Desmodium triflorum

Crotalaria retusa

Lablab purpureus

Lonchocarpus benthamianus

Moghania strobilifera

FABACEAE (continued)

Sabinea carinalis Gr.*

Local names: Carib Wood, 'Bois Caraïbe' (Dominica).

Indigenous to Dominica, and found in dry coastal areas along the leeward coast especially. It is Dominica's new national flower.

A low, spreading tree with a feathery habit, it blooms in early May. It has finely bipinnate leaves. The flowers appear in clusters along the branches, and are a brilliant crimson. Without the presence of the leaves, the tree has a very oriental appearance with its mass of blossom along the branches. The fruit is a long pod and the seeds propagate readily.

FLACOURTIACEAE

Homalium racemosum Jacq.*

Local name: 'Acomat' (Dominica, Fr. West Indies).

Found in the West Indies at middle elevations — usually in windbreaks in Dominica.

A shrub or tree which may grow to 16 m high. The leaves are elliptical and papery in texture. The flowers are small and white. The tree has a tall thin trunk and a spreading habit. It flowers May to June. The leaves turn yellow when about to fall.

The wood is often used in furniture or house building.

GENTIANACEAE

Leiphamos aphylla (Jacq.) Gilg. in Engl. & Prante (*Voyria aphylla*)*

Found from Costa Rica to Paraguay and in the West Indies, it is not commonly found in Dominica.

This small plant is saprophytic, having no chlorophyll. It lives on dead or decaying organic matter. The stem is approximately 8–18 cm long and is whitish with tiny scale-like leaves ascending towards the yellow flower.

Calolisianthus frigidus (Sw.) Gilg*

Found in Guadeloupe, Dominica and St. Vincent (xvii) in volcanic cones at high elevations.

This unusual showy plant has a shrubby habit, a yellow-green corolla, and long trailing tetragonal branches.

GESNERIACEAE

Alloplectus cristatus (L.) Mart.*

Local name: 'Fuchsia Sauvage' (Fr. West Indies).

Found at upper elevations, in moist locations.

A straggling climber with opposite leaves. The flower calyx is a brilliant scarlet, and the flower is yellow and tubular.

48

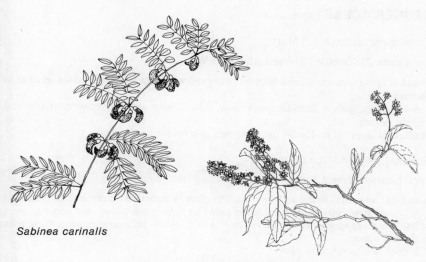

Sabinea carinalis

Homalium racemosum

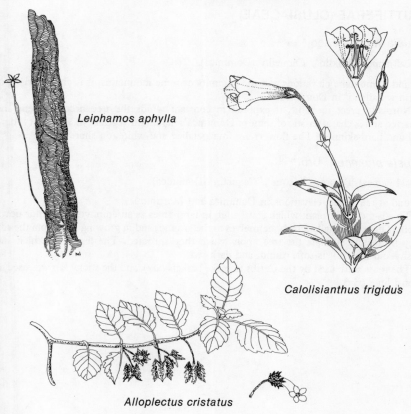

Leiphamos aphylla

Calolisianthus frigidus

Alloplectus cristatus

49

GESNERIACEAE (continued)

Episcia melittifolia (L.) Mart.*

Local name: 'L'Oiseille', 'Herbe à Meil' (Fr. West Indies).

Found at upper wet-area elevations, near rivers and waterfalls and roadbanks. Common in Dominica.

A low plant with a flaccid, square stem. The flowers are mauve, campanulate and slightly velvety.

Medicinal uses   The Caribs use this plant as a tea for colds (xix).

Gloxinia perennis (L.) Mart.*

Local names: 'Herbe à Veuve', 'Gueule de Loup' (Fr. West Indies).

Native of Colombia, Brazil and Peru (i), this plant is sometimes cultivated in gardens.

A fleshy herbaceous plant about 30 cm tall. The flowers are velvety and occur on erect stalks. The plant spreads easily, flowering from October to February and then dying back in the dry season.

It does well as a border plant.

GUTTIFERAE (CLUSIACEAE)

Clusia venosa Jacq.*

Local names: 'Kaklin', 'Caquelin' (Dominica).

Found at montane elevations and the summits of some mountains, it is characteristic of 'elfin woodland' in Dominica.

Normally erect in habit, if exposed to constant winds the tree becomes bent and dwarfed and is characteristic of some of Dominica's mountain vegetation. The fruit is dark red and hard-skinned. The flowers are four-petalled and white, on short stems.

Clusia plukenetti Urb.*

Local names: 'Pomme Chicque', 'Caquelin' (Dominica).

Found at rain forest elevations in Dominica and Martinique.

This is a curious plant which starts high in large trees as an epiphyte and sends down roots. These eventually attach themselves to the ground, and in growing now from the soil they ultimately strangle the tree upon which they sprouted. The leaves are thick and fleshy and the fruit is soft, round, and dark red.

The roots were used by the Caribs to make baskets (xix), and the sticky sap was used as a bird lime (xix).

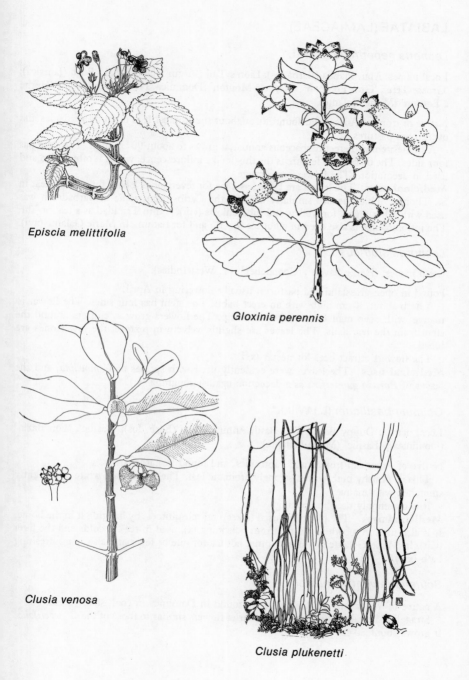

Episcia melittifolia

Gloxinia perennis

Clusia venosa

Clusia plukenetti

LABIATAE (LAMIACEAE)

Leonotis nepetifolia (L.) Ait.*

Local names: Man Piabba (Barbados); Lion's Tail (Virgin Islands); Bald Bush (Jamaica); 'Grosse Tête', 'Gros Pompon', 'Herbe à Mouton' (Dominica, Fr. West Indies); 'Z'Herbe à Bouton' (Marie Galante).

Native of tropical Africa, and found throughout the tropics. It is often by roadsides and overgrown pastures.
   A single-stemmed erect herbaceous annual, it grows to about 90 cm high. The stem has four sides. The distinctive feature is the thistle-like inflorescence, which is often dried and used in decoration. The flowers are orange.
Medicinal uses   The leaves are boiled as a tea for fevers or as a bath for prickly heat in St. Thomas. This was also similarly used by the Caribs (xix), while in Barbados it was used with other plants for worms and tuberculosis (viii), and in Trinidad as a tea for 'flu. The inflorescence is also used in a hot tea for 'flu and for bathing in Marie Galante (xvi).

Leonorus sibiricus L.*

Local names: 'Herbe Savon', 'Chandelier' (Fr. West Indies).

Found in open grassland and pastures, usually flowering in April.
   A herb about 30 cm high with an erect habit. The stem has four sides. The flower is mauve, with deep mauve markings on the lip. The flowers grow in clusters around the stem from the leaf axils. The leaves are slightly velvety in texture, the lower ones are lobed.
   The flowers attract bees for nectar (xxi).
Medicinal uses   The leaves were evidently used with species of *Commelina*, and the leaves of *Bursera gummifera* as a decoction against coughing (xi).

Ocimum basilicum (L.) Willd.*

Local names: Duppy Basil, Wild Basil, Annual Weed (Barbados); 'Basilic', 'Fon Bazin' (Dominica); 'Basilic' (Marie Galante).

Native of American tropics and subtropics, and found in gardens.
   This is a bushy herb with a strongly aromatic leaf. The flowers are white and slightly splotched with mauve.
   It is commonly used in cooking.
Medicinal uses   The plant was used to keep off mosquitoes, by hanging it in the home. It is also used in a bath to cleanse from poisoning (x), used in tea for colds and the liver (viii). In Marie Galante it was taken in a hot tea for bile or fever and also in a refreshing bath (xvi).

Scutellaria ventenatii Hook.*

A native of Venezuela and Colombia (i), found in Dominica in cool, shady areas.
   Straggling in habit, this plant bears scarlet flowers similar to those of the *L. nepetifolia*. It grows about 30 cm high.

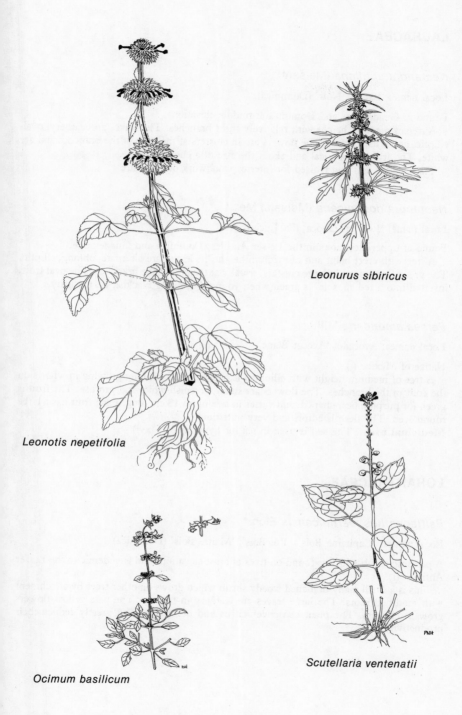

Leonurus sibiricus

Leonotis nepetifolia

Ocimum basilicum

Scutellaria ventenatii

LAURACEAE

Nectandra antillana (Meissn)*

Local name: 'Laurier Noir' (Dominica).

Found in Guadeloupe and Dominica at middle elevations.
  A tree of medium height with relatively short branches. The leaves are leathery, oval-elliptic, and a dark green. The flowers are in clusters at the ends of the branches and are white. The fruit is spherical and about the size of a cherry.
  The wood is soft and is used for interior woodwork or furniture.

Nectandra dominicana (Meissn.) Mez.*

Local name: 'Laurier Zaboca' (St. Lucia).

Found at upper elevations in the Lesser Antilles, Dominica and Guadeloupe.
  A tree with erect habit and of medium height, its leaves are alternate, oblong, elliptic. The green flowers are in large clusters, usually at the end of the branches. The oval fruit, inserted into a red capsule, is green when young, becoming blackish on maturity.

Persea americana Mill.

Local names: Avocado, 'Avocat Blanc'.

Native of Mexico (i).
  A tree of medium height with elliptical leaves up to 18 cm long growing in whorls on the ends of the branches. The flowers are small, greenish-white and profuse. The fruit is green (or purple), pear-shaped, and varies in size up to 18 cm long. Each fruit has a large round seed. The flesh is edible and very nutritious.
  Medicinal uses The leaf is used in tea for hypertension (xvi).

LORANTHACEAE

Psittacanthus martinicensis Eich.*

Local names: 'Capitaine Bois', 'Roi Bois', 'Maitre Bois' (Dominica).

A parasitic shrub which is found on trees at most elevations and is endemic to the Lesser Antilles. (i).
  This is a small many-branched woody shrub which grows on other trees by attachment with modified roots. The large leaves are leathery in texture. The pale yellow flowers grow in clusters. The plant can cover citrus and other trees and greatly reduce their productivity.

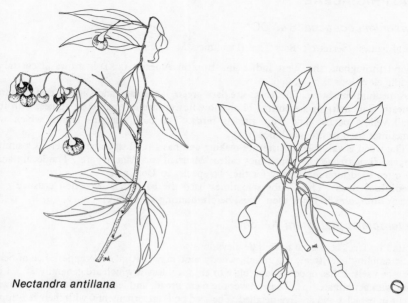

Nectandra antillana

Nectandra dominicana

Psittacanthus martinicensis

MALPHIGIACEAE

Byrsonima coriacea (Sw.) DC*

Local names: 'Serrette'; 'Bois Tan' (Dominica, St. Lucia).

Found throughout the West Indies and tropical America (xix). It grows at coastal to middle elevations.

A medium-sized tree. The leaves are dark green, lanceolate, and slightly rust-coloured beneath. The margins are entire. The yellow flowers are in small erect spikes, with five small yellow petals. The fruit attracts birds. The bark contains tannin which was sometimes used to prepare hides (xix).

The wood is valuable in furniture making and has a reddish colour. There is a similar species, *B. martinicensis*, sometimes called 'Mauricif' or 'Mauricypre'. The local names are often used interchangeably for the two species in Dominica.

Medicinal uses The Caribs sometimes used the bark in a decoction crushed with *Hyptis atrorubens* to bathe their dogs, before hunting agoui (xix).

Heteropteris platyptera A.P.deC.*

Found on trees at coastal to middle elevations.

Scrambling over trees, this tough, woody vine may first give the appearance of being the tree itself. It has opposite lanceolate to elliptical leaves which are generally 20–21 cm long with acuminate tips. These leaves are 6 cm broad, and leathery. The yellow flowers occur in panicles and are five-petalled. The seed pods are ornamental with their red-tipped wings, and these usually emerge on the lower stalk in groups of two or three. The seed is encased in the lower margin at the base of the pod. Blooms from April to May.

Stigmaphyllon cordifolium Niedenzu*

Local name: 'Aile à Ravet' (Marie Galante).

Found at low coastal elevations in the West Indies.

This vine has leaves which are glossy in texture above, but with a silky down underneath. The yellow flowers occur in small clusters along the stem and have five small, distinct petals. The seeds are winged.

A similar species, *Stigmaphyllon puberum* (L. C. Richard) A. Juss, was used by the Caribs as roping to tie the rods or poles of their dwellings (xix).

Medicinal uses Used in teas to counteract 'flu (xvi), and also to correct irregularities in menstruation, and in baths (xvi).

MALVACEAE

Abutilon indicum (L.) Sweet*

Local name: 'Guimauve' (Guadeloupe and Marie Galante).

Found in tropical America (i), it grows in open pastures at low elevations.

This plant grows about 30–40 cm high, and has alternate leaves which are ovate with slightly serrated margins. The flowers occur at the base of the leaf axils on long stalks. These are yellow and have five petals. The carpels have many sections in a circular arrangement, and are dark brown when dry.

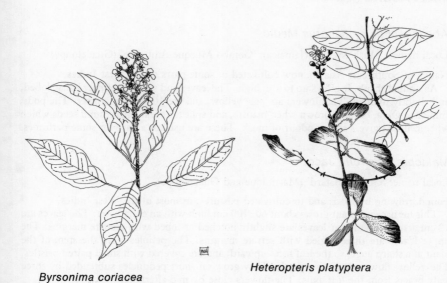

Byrsonima coriacea

Heteropteris platyptera

Stigmaphyllon cordifolium

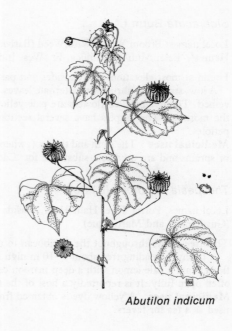

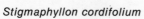

Abutilon indicum

MALVACEAE (continued)

Abelmoschus moschatus Medic.*

Local names: Musk Mallow (Jamaica); 'Gombo Musque Ambrette' (Guadeloupe).

Native of South-east Asia (i), now cultivated in some parts of the West Indies.
 An annual shrub. It grows up to 1 m high. The leaves and stems are finely hairy, lobed, and slightly toothed. The flowers are pale yellow, and are 10 cm in diameter. The pods, similar to Okra, become brown when mature, and split to reveal small round seeds which when crushed give a strong odour of musk. These are used as a base for some perfumes.

Malachra alceifolia Jacq.*

Local name: 'Gombo Bâtard' (Martinique and Guadeloupe).

Found growing by roads and in cultivated pastures, in most of the West Indies.
 This herbaceous plant grows about 60–100 cm high with an erect habit. The leaves are 12 cm across. The lower leaves are slightly notched or lobed with crenate margins. The upper leaves are three-angled with serrate margins. The petioles leave the stem of the plant at a sharp angle so the leaf faces upward, and are covered with small paired bristles. The yellow flowers have five petals. They grow on short peduncles subtended by three leafy bracts from the leaf axils. The flowers close by mid-afternoon.

Sida acuta Burm f.*

Local names: Broom Weed, Wire Weed (Barbados, Jamaica); 'Balier Savane', 'Balai Onze Heures', 'Balai Midi' (Dominica, Fr. West Indies).

Found at most elevations on roadsides and pastures.
 A low, straggling shrub with alternate leaves. These are slightly toothed and noticeably veined. The five-petalled flowers are pale yellow and 1–2 cm across. They open only in the morning. The carpels have several sections. Flowers occur at the axils of the leaf petioles.
Medicinal uses The stem and the root, when pounded, are used as a poultice on strains or sprains and are used as a substitute for 'Cachima', *Annona reticulata*.

Thespesia populnea (L.) Sol.*

Local names: Portia Tree, Haiti Haiti, Seaside Mahoe (St. Thomas); 'Calpata', 'Calfata' (Guadeloupe and Martinique).

This tree grows throughout the Caribbean in the coastal regions.
 A bushy, spreading tree about 9–10 m high, the leaves are cordate and alternate. The flowers are a pale lemon with a deep maroon centre. They have five petals which do not often open fully. It is reportedly a host of the Cotton Stainer insect.
Medicinal uses A yellow dye is obtained from the unripe fruit. The leaves have been used as a tea for fevers.

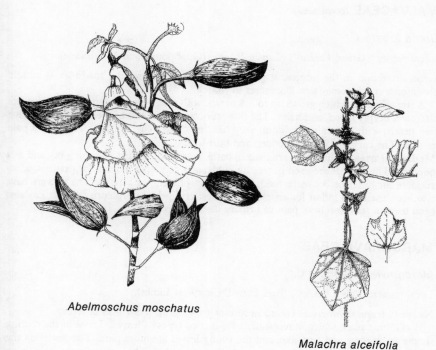

Abelmoschus moschatus

Malachra alceifolia

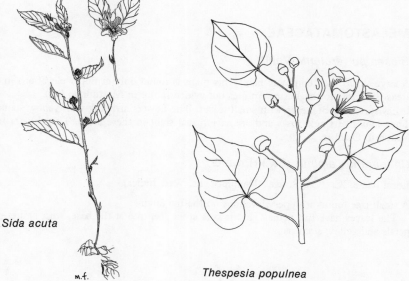

Sida acuta

M.f.

Thespesia populnea

MALVACEAE (continued)

Urena lobata L.*

Local names: 'Grand Cousin', 'Cousin Rouge' (Guadeloupe and Martinique).

Generally found in the tropics. It grows in cultivated pastures, by roadsides at middle elevations, and blooms from December to March.

A straggly herb which grows up to 1.5 m tall, with tough woody stems. The leaves are alternate, deeply lobed, and hairy. The flowers occur in the axils of the leaves. They have five petals, with a staminal column, very like the cultivated Hibiscus. They are a pale purple. The carpel is in five sections and burr-like.

Medicinal uses   The leaves are used in baths to refresh, against poisoning (x), and as a poultice. The flowers are used in teas for colds, as a gargle, as a drink for gastritis, as an antidote for bitter or corrosive foods, and against phlegm, or catarrh. The flowers have also been used in a lotion for eruptive maladies, erysipelas and pleurisy (xi). It has also been used in teas to relieve pain of urinary disorders (xi).

MARCGRAVIACEAE

Marcgravia umbellata L.*

Local names: 'Bois Couilles', 'Bois Petard' (Fr. West Indies).

Native to tropical America. Found in forests growing up the trunks of trees.

A climbing plant which is remarkable for its two types of leaves. Those of the rooting shoots are cordate and fern-like, and the young leaves are often pink. The leaves on the climbing stems are linear, alternate, with fleshy petioles. The green flower head is an umbel, and the flowers (shown here in bud) will turn over to face upwards when open. The central parts are nectaries.

MELASTOMATACEAE

Blakea pulverulenta Vahl*

A large shrub with epiphytic habit, this plant is found only at upper elevations in wet areas. Endemic to the Lesser Antilles but evidently not in Martinique (xii).

It has spreading branches with small leaves. The flowers are pink with yellow stamens. They look like single roses and are an unusual sight in these areas. They bloom from March to October.

Charianthus alpinus (Sw.) Howard*

Local name: 'Cré Cré Rouge' (Dominica, Fr. West Indies).

A small tree found at upper elevations on the mountains.

The leaves have five veins. The flowers are a deep red at the base, with lighter-pink petals and yellow stamens.

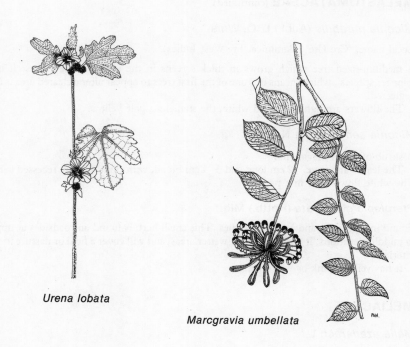

Urena lobata

Marcgravia umbellata

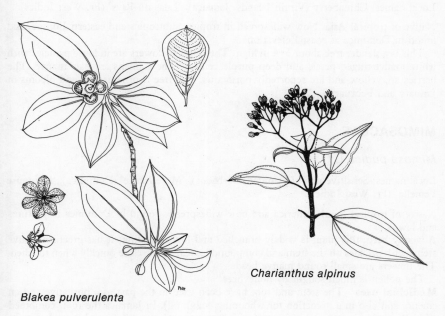

Blakea pulverulenta

Charianthus alpinus

MELASTOMATACEAE (continued)

Miconia mirabilis (Aubl.) L. O. Wms.*

Local name: 'Cré Cré' (Dominica, Fr. West Indies).

A medium-sized tree which grows in thick screens in the upper, wetter areas. It is a 'pioneer species', which means it is one of the first trees to appear after a cleared area is left neglected.

The flowers are light pink and white, the stamens a pale yellow.

Miconia semicrenata Naud.*

A shrub of upper elevations.

The leaves are ovate, 12 cm long and 3–4 cm broad, with three slightly recessed veins. The white flowers are in a panicle.

Pterolepis glomerata (Rottb.) Miq.*

Found in Dominica and Fr. West Indies. This small herb is found on roadsides at upper to middle elevations. It seems to prefer wetter areas, and will cover a field or pasture in the interior.

It has yellow, pink or white petals.

MELIACEAE

Melia azedarach L.*

Local names: Chinaberry (Virgin Islands, Jamaica); 'Lilas du Pays' (Fr. West Indies).

Native of tropical Asia. Now widespread in tropics, subtropics and eastern Europe, and found in Dominica at coastal elevations.

This is a slender tree about 3–4 m high. The fragrant flowers are in loose panicles with white mauve-tinged petals and deep purple centres. The stamens are pale mauve. The berries are yellow and are reportedly poisonous. The tree is ornamental, and blooms in January and February.

MIMOSACEAE

Mimosa pudica L.*

Local names: Sensitive plant (Trinidad, Barbados); 'Mese Marie' (Dominica); 'Honteuse Femelle' (Fr. West Indies).

Native of tropical South America and now widespread. Found in Dominica in pastures and lawns.

A woody herb, this plant is widely branched and low spreading. It has greenish-mauve stems with prickles on the stem and compound leaves. These close quickly when touched. The flowers are small round mauve heads.

The pollen is much sought by honeybees.

Medicinal uses The stem and root have been used in the past as a purgative and an emetic, and also in a decoction for whooping cough (xi). In Jamaica the root, combined with *Desmodium* spp., and *Achyranthes indica* was used as a tea for colds, and also for stomach and urinary disorders. In Marie Galante it is taken in a cooling tea (xvi).

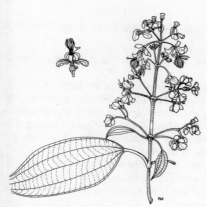

Miconia mirabilis

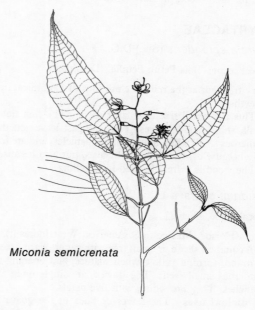

Miconia semicrenata

Pterolepis glomerata

MORACEAE

Cecropia peltata L.

Local names: Trumpet Tree (Trinidad and Tobago, Barbados); 'Bois Canon' (Dominica, St. Lucia, Fr. West Indies).

Found from Mexico to Colombia and throughout the West Indies (i), in secondary forests at middle elevations.

A slender tree which may grow up to 25 m. The palmate leaves are large and deeply lobed into seven to eleven sections. These are alternate and form whorled clusters at the ends of the branches. Their undersides are a distinctive white and are easily seen when blown by wind. The trunk is smooth and grey and when young may have a tendency to develop prop roots. The grey-brown fruit is a cluster of cylindrical spikes. The male and female flowers are on different trees, at the base of the leaves.

The wood is used primarily in raft making, and for boards or palings.

Medicinal uses   The leaves are made into a tea in Guyana to clean the kidneys and also as a soporific. It was popularly used in Barbados for kidney troubles (viii), for diabetes, and for high blood pressure (viii).

MYRTACEAE

Myrcia splendens (Sw.) D.C.

Local name: 'Bois Petite Feuille' (Dominica).

Very frequent at the middle elevations of Dominica, in windbreaks and secondary forest growths.

This is a small tree or a large shrub. The first noticeable feature is its profusion of small, shiny, single leaves. The youngest leaves on the branches are a pale pink. The numerous small white flowers are in panicles, and are fragrant. It is a good shade tree, not only because of its many small leaves, but also because it is evergreen.

The wood is primarily used for kindling.

Psidium guajava L.

Local name: Guava.

From Florida, Mexico to S. America, West Indies (i).

A common shrub or small tree with rough, light-green elliptical leaves. The edible fruit is round, turning yellow when mature, and opening to reveal a bright pink pulp with numerous small seeds. The flowers are single or in a small cluster, on the ends of the branches. They are white, with five petals.

Medicinal uses   The leaves or buds of the guava are used medicinally in a tea for dysentery (xi), stomach-ache, worms in children (x) and for diarrhoea. For the latter they are taken with the leaves of *Citharexylum spinosum*.

Pimenta racemosa (Mill) J. W. Moore

Local names: Bay; 'Bois d'Inde'.

Native of northern S. America and West Indies (i).

A cultivated tree which grows up to 10 m high. The bark is smooth and grey, and the branches tend to grow upward, giving a compact appearance. The elliptic leaves are glossy and the small flowers are white.

Medicinal uses   Used in baths to refresh and relax, taken on three consecutive days. A small leaf is often used in hot teas to counteract chills.

Melia azedarach

Mimosa pudica

MYRTACEAE (continued)

Syzgium aromaticum (L.) Merr. & Perry*

Local names: Clove Tree; 'Klujooff' (Dominica).

From Zanzibar and Madagascar and found in some gardens in Dominica. Grows best at middle elevations.
   A small, compact, upright tree. The outer leaves are reddish when young. They are lanceolate and shiny. The flowers are in bright-pink clusters at the ends of the branches, and have numerous feathery, white stamens. The clove itself is the receptacle of the flower which is picked and then dried. It is propagated by seeds.
Medicinal uses Commonly used to combat toothache.

Syzgium jambos (L.) Alston*

Local names: Rose Apple, 'Pomme Rose' (Dominica, West Indies).

Native of the Indo-Malaysian and Pacific regions (i) and now widely cultivated in both tropics. Found in Dominica in windbreaks.
   A medium-sized evergreen tree with a spreading habit. The leaves are dark glossy green. The flowers are pale ivory. The stamens are the showy part of the flower, while the petals, which are round and waxy, are not much in evidence. The fruit is ivory to pale pink and strongly fragrant.
   The fruit is used in India and Malaya as a preserve. In Dominica, the tree was introduced to shade coffee and cocoa plantations.

NYCTAGINACEAE

Boerhavia coccinea L.*

Local names: Hog Weed (Barbados); 'Patagon Rouge' (Fr. West Indies).

Found from Mexico to Argentina (xi), this weed grows in dry coastal areas.
   A small, thin weed with opposite leaves. These are ovate, and found at the base of the stem. The mauve-red flowers are extremely small and the seeds are ovoid. (This family is perhaps best known for the genus *Bougainvillea*, which is common throughout the West Indies.)

OCHNACEAE

Sauvagesia erecta L.*

Local names: 'Petit Mayoc', 'Petit Manioc' (Dominica).

Native to the tropics and subtropics (i), this small herb is usually found at middle to upper elevations.
   It is a small bushy herb, erect in habit, with alternate leaves, which are lanceolate and 3 cm long. Small hairy stipules occur at the base of each leaf. The flowers are white and insignificant, with five stamens. The fruit is a capsule.
Medicinal uses Primarily it is used as a tea with *Polygala paniculata* for bad colds. It is also used as a bath after parturition (xix).

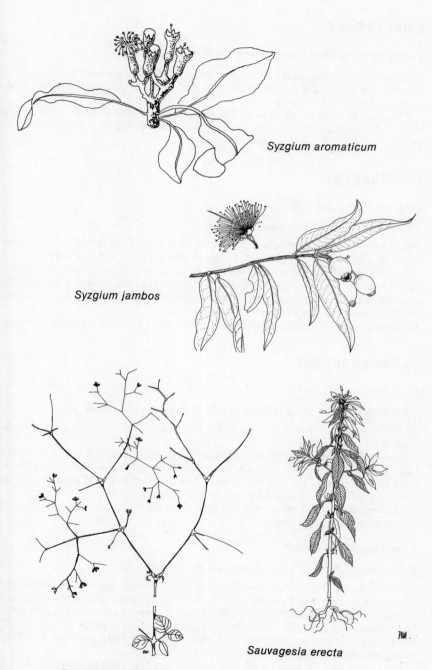

Syzgium aromaticum

Syzygium jambos

Boerhavia coccinea

Sauvagesia erecta

ONAGRACEAE

Ludwigia octovalvis (Jacq.) Raven*

A low herbaceous plant found in pastures and roadsides at upper elevations. Generally found in the tropics and subtropics (i). The plant seems to prefer damp places, and is found in ditches and by ponds.

This plant is about 30–60 cm high. The leaves are linear, and 6–8 cm long. The lower sides of the leaves are finely hairy. The four pale-yellow petals are 1 cm long and 5 mm broad.

OXALIDACEAE

Oxalis corymbosa D.C.*

Local name: 'Malgoj' (Dominica).

Found from Florida to South America, and the West Indies, and considered an ornamental in the Old World (i). It is found in little clusters on paths and roadways.

This small herb has leaves and flowers rising straight from the ground to a height of 10–12.5 cm. The leaves are trifoliate and fold up at night or in hot sun. The mauve bell-shaped flowers have five petals and grow on long slender peduncles in clusters of three or four. The stem and stalks contain an acrid juice.

Medicinal uses   The leaves are chewed, or brewed in a cooling tea for sore throats.

PASSIFLORACEAE

Passiflora foetida L.*

Local names: 'Marie Gouja' (Dominica); Love-in-a-Mist (Barbados).

Found in hedges, thickets and along paths in coastal and low areas. A pantropical weed. (i).

This twining vine has three-lobed leaves, which grow alternately. A tendril grows at the base of each petiole. Small hairs occur on the leaves and stems. The flowers are surrounded by filaments. They are purple and white, 7–12 cm long, and are on a long stem. The small fruit is also enclosed by very hairy filaments.

Medicinal uses   After drying it makes an effective tea for sore throats. In Jamaica, this tea was given for kidney trouble (iv).

Passiflora laurifolia L.*

Local name: 'Pomme de Liane' (Dominica, Fr. West Indies).

Native of tropical South America and the West Indies, this vine has been introduced to the Old World (i). Found in Dominica at middle elevations in thickets, on trees and scrub.

This is a heavy vine which climbs into the trees and forms a thick mat. The leaves are 10 cm long, single, alternate, and shiny dark green. The stem is tough and grooved. The sweet-scented flowers are ivory-coloured with purple and ivory stamens. The fruit is oblong or ovoid, turning yellow when mature.

The stem is used for basket work. It forms the frame of round or shallow baskets and other utilitarian work baskets. It dries hard and strong, and then is woven in lateral strips. Finally strips of bamboo may be added from the middle to the top of the basket.

Ludwigia octovalvis

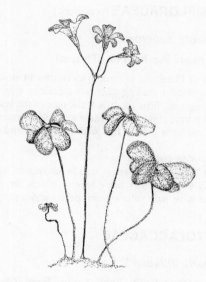

Oxalis corymbosa

Passiflora foetida

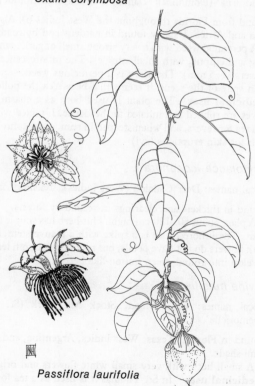

Passiflora laurifolia

PASSIFLORACEAE (continued)

Passiflora suberosa L.*

Local name: Pap Bush (St. Thomas).

Found in Dominica at middle elevations in shady thickets.
   This slender trailing vine has a delicate habit. The simple alternate leaves are about 8 cm long with little tendrils at the base of the leaf stem. The small, green-petalled flowers have a purple ring in the middle and yellow-green stamens, stigmas and style. The fruit is small and light green. The stem is reddish and very slightly hairy.

Passiflora rubra L.*

A vine found in thickets at middle elevations in Dominica.
   This vine has soft hairy leaves, which are two-lobed and cordate at the base. The flowers have pale yellow-green petals and a purplish corona.

PHYTOLACCACEAE

Petiveria alliacea L.*

Local names: Gully Root, Garlic Weed (Barbados); Guinea Hen Weed (Jamaica); 'Kudjuruk' (Dominica); 'Danday', 'Arada' (Martinique).

Found from Florida throughout the West Indies (i). Also found in some parts of tropical Asia and Africa. Usually found in gardens and by roadsides at lower elevations.
   A perennial weed with a very strong smell of garlic when crushed. The leaves are 10 cm long and elliptic, with small stipules. The inflorescence is a spike with very small white flowers (4–5 mm). The plant is a pernicious weed.
   In spite of the scent, it seems that bees seek the pollen in the early morning (xxi).
Medicinal uses The plant is used both as a charm and a medicine. In Jamaica the leaves are rubbed and inhaled for headache. If mixed with *Eryngium foetidum* it was used in a tea for fevers, and when steeped in rum, as an aphrodisiac (v). The root is steeped in a bath for skin eruptions (xvi).

Phytolacca icosandra L.

Local names: Deer Callalou (West Indies); 'Raisin d'Amerique' (Fr. West Indies).

Found in thickets and clearings in secondary forests.
   A plant about 60–120 cm high. This herb has simple light-green leaves, alternate on the stem. The inflorescence is a spike with a mauve stem, and small mauve peduncles. The pink flowers do not have petals, but rounded perianth leaves, which are five-parted. These develop into round green button-like berries.

Rivina humilis L.*

Local names: Cat's Blood, Stock Ma Hork (St. Thomas); Bloodberry (Jamaica) 'Demoiselle'.

Found in Florida, Texas, West Indies, Argentina, and naturalized in tropical Asia (i), in semi-shade.
   A small herb with very small white flowers and bright red berries.
Medicinal uses In St. Thomas it is used as a tea for diarrhoea, and in Dominica, the Caribs used it as a medicinal tea (xix).

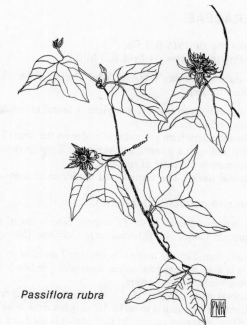

Passiflora rubra

Passiflora suberosa

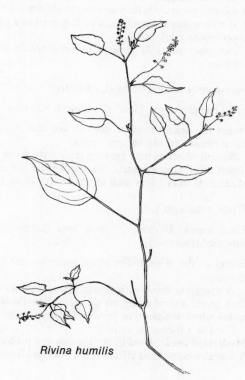

Petiveria alliacea

Rivina humilis

71

PIPERACEAE

Lepianthes peltata (L.) Raj.*
(syn. *Pothomorphe peltata* (L.) Miq.)

Local names: Monkey's Hand; 'Mal Dormi' (Dominica); 'Mal Nommé', 'Mal Tête' (Fr. West Indies).

From the American tropics, this plant is found on roadsides, in damp places, usually at lower to middle elevations.

A shrubby herb with pronounced nodes on the stem. The leaves are peltate with a blunt acute tip and with a pronounced venation. They grow up to 15 cm or more across. Each inflorescence is a cluster of spikes.

Medicinal uses   The Caribs used the leaves as a poultice for headaches (xix).

Peperomia pellucida (L.) Kunth.*

Local names: Shine Bush (Tobago); Silver Bush (Barbados); 'Z'Herbe Couesse' (Dominica); 'Coquelian' (Martinique); 'Cochlaia' (Marie Galante).

Generally found in the tropics in neglected areas or in gardens.

This is a small green succulent herb with cordate leaves, and with terminal spikes of minute greenish-white flowers.

Medicinal uses   In Trinidad it is used as a cooling tea and as a tea for colds (xxx). In St. Thomas it has similar uses. In Africa it is a remedy against convulsions (vii), and is said to be effective against cancer. In Barbados it is used as a diuretic for kidney ailments (viii). It was also given there for marasmus in children. The Caribs used to make a poultice from it to relieve sore throats (xvi, xix). It is used as a cooling tea. Used in Marie Galante for hypertension (xvi).

Peperomia pellucida appears to be generally used for asthma and for pulmonary troubles.

Peperomia rotundifolia (L.) Kunth.*

Local names: 'Giron Fleur' (Dominica); 'Giroflée' (Martinique Guadeloupe).

Found throughout tropical America and also tropical Africa (i). Found in Dominica in damp places, usually on tree trunks.

A small prostrate herb growing on tree trunks or rocks. Vivid green, with small, round almost membraneous petals, the whole plant is flaccid and wilts quickly when picked.

Medicinal uses   It is used as a tea for colds in Dominica.

Piper amalago L.*

Local names: Blackwattle, Sout Sout (Jamaica); Joint Wood (Virgin Islands); 'Mal Estomac' (Dominica).

Found in the West Indies along roadsides and banks usually in semi-shade, at most elevations.

A straggling shrub with woody stems, it grows up to 2 m high. The leaves are ovate, dark green, and alternate on the stem. The flowers are very tiny and in dense, fleshy spikes which are green or brown. These spikes occur at the base of the petiole.

Used as a fodder for rabbits.

Medicinal uses   Used in St. Thomas in a tea for colds and fever, and for constipation. It was also considered effective as a sedative. It was used by Caribs in ritual baths (xix).

72

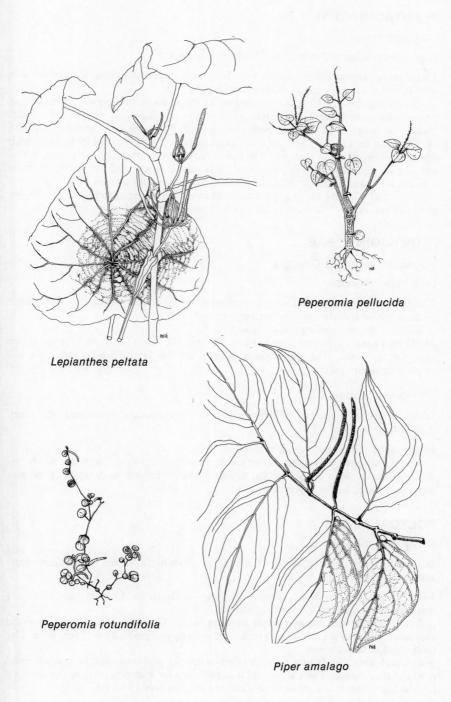

Lepianthes peltata

Peperomia pellucida

Peperomia rotundifolia

Piper amalago

73

PLANTAGINACEAE

Plantago major L.*

Local names: 'Plantain' (Dominica); Millet (Martinique).

This plant has a cosmopolitan distribution. It is found at most elevations, often growing in poor soil conditions, in cracks, rubble and on old paving.

A small perennial herb. It is identified by its basal rosette of leaves. The stalk bears insignificant green flowers. The leaves are spatulate, and somewhat wavy at the margin.
Medicinal uses This plant has been used successfully for eye inflammation (xvi). The Caribs used it for this (xix), and Beckwith (iv) stated that it was also used in the eighteenth century as a decoction or eyewash. The leaves have been used as a tea with mint, thyme and salt to counteract shock. Europeans in South America used the leaves to dress wounds, sores and ulcers. In Africa the leaf juice was used for malaria (iv).

It is used as a cooling tea in Dominica. In Martinique it is used as a hot tea to bring on menstruation and for other complaints such as wind, or stomach pains (iv).

PLUMBAGINACEAE

Plumbago capensis (Thunb.)

Local name: Plumbago.

The cultivated plumbago is native to tropical Africa (i) and is common in West Indian gardens, where it makes an effective border plant.

This is a shrub with large, blue flowers.
Medicinal uses It has been used in a cooling tea. The leaves and roots were sometimes used for abortions. The **deleterious effects** are blistering, heart contraction and respiratory failure (viii).

Plumbago scandens L.*

Local names: 'Sinapisme' (Martinique and Guadeloupe); 'Moutarde du Pays' (Guadeloupe).

Found at lower elevations along roadsides.

This is a straggling shrub with ovate leaves about 5–6 cm long. The white flowers are about 1.5 cm across with five petals, subtended by sticky and hairy sepals. It is quite commonly found in the lower elevations.

POLYGALACEAE

Polygala paniculata L.*

Local names: 'Estroi Fragile', 'Essence Fragile' (Dominica); 'Herbe à Lait (Martinique and Guadeloupe).

Found from Mexico to Brazil, in the West Indies, Malaysia and Oceania (i); in Dominica it is common in gardens at middle elevations.

A small annual with feathery simple alternate leaves. The small, pale-mauve flowers are on a spike and measure 2–3 mm. The herb is delicate and grows up to 30 cm high. The roots smell of wintergreen.
Medicinal uses Used in a tea with other plants for problems such as 'crazy people' (fits) and rheumatism. The Caribs used it in baths and for sponging off their canoes (xix). Crushed in water, it is used to wash a baby to help its bones knit (x).

74

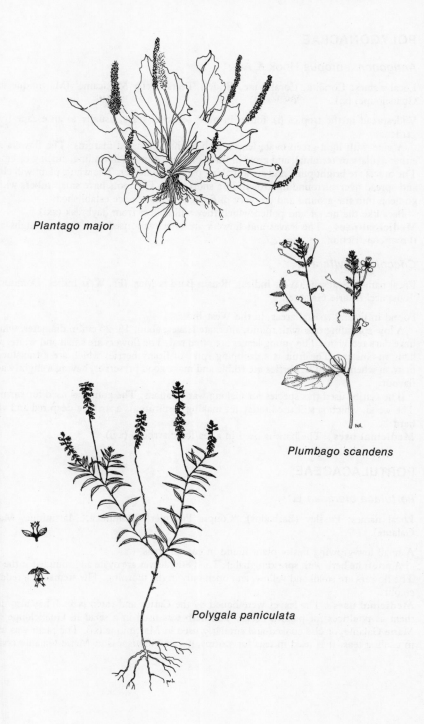

Plantago major

Plumbago scandens

Polygala paniculata

POLYGONACEAE

Antigonon leptopus Hook & Arn.*

Local names: Coralita, Coral Vine, Bride's Tears 'Belle Mexicaine' (Martinique and Guadeloupe) (xi).

Widespread in the tropics (i), found in most sea-level areas, usually as an escape from gardens.
  A vine with light-green ovate leaves with slightly crenated margins. The flowers are either axillary or terminal, and grow in three-pronged clusters in trailing, curling racemes. The bracts are bright pink, and are a striking part of the flower. The whole plant will cling and spread over surrounding growth in a short time. The roots have small tubers which go deep into the ground and are very difficult to remove once established.
  Bees like the nectar and pollen which they will collect from daybreak (xxi).
Medicinal uses The leaves and flowers are effectively used in a tea for coughs and throat constriction.

Coccoloba uvifera L.

Local names: Seagrape (West Indies); 'Raisin Bord la Mer' (Fr. West Indies, Dominica); 'Raisinier' (Marie Galante).

Found in littoral coastal areas, in the West Indies.
  A low spreading tree with round, alternate leaves, about 15–20 cm in diameter, which have dark red veins. The young leaves are often red. The flowers are small and white, and hang in clusters. The fruit is a dropping spike of fleshy berries which are a translucent maroon when ripe. The berries are edible and make good preserves, having a slightly acid flavour.
  The Caribs used this species for making war weapons. The gum was used for varnish. The wood, which is still used today for making furniture, is a striking deep red and very hard.
Medicinal uses The bark is used in a tea for diarrhoea (xvi).

PORTULACACEAE

Portulaca oleracea L.*

Local names: Pussley (Barbados); 'Coupie Pourpier' (Dominica, Martinique, Marie Galante).

A small low-growing fleshy plant found in coastal areas.
  A prostrate herb with spreading habit. The fleshy leaves are ovate and rounded at the tip. The flowers are small and yellow, and open only in the morning. The stems are a reddish colour.
Medicinal uses The leaves were boiled by the Caribs and eaten (xix). They also used them as poultices for painful backs. The herb was used in a salad in Guadeloupe and Marie Galante, or else cooked and similarly used in Martinique (xi). The plant was used in cooling teas. It is used in teas for worms, gas or biliousness in Marie Galante (xvi).

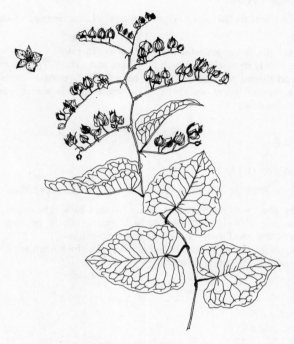

Antigonon leptopus

Portulaca oleracea

RANUNCULACEAE

Clematis dioica L.*

Local names: Wild Clematis (Jamaica); 'Vigne Sauvage', 'Liane Serpent', Liane à Crabes' (Fr. West Indies).

A vine found at middle elevations which spreads in windbreaks.

    This vine has a woody stem which is slightly twisted and ribbed. The smooth leaves are ovate, almost heart-shaped and are palmately veined. The younger leaves are deeply serrated, but the mature leaves are entire. The white flowers are in panicles. This illustration shows the fruits.

RHAMNACEAE

Gouania lupuloides (L.) Urb.*

Local names: White Root (St. Thomas); 'Liane Savon' (Guadeloupe, Martinique).

A woody climbing plant with angular stem. It has fissured bark. The opposite leaves are ovate, rounded at base with serrated margins with tendrils on the petioles. The inflorescence is racemose and the flowers are a light greenish-yellow (i).

Medicinal uses It is used in Jamaica as a dentrifice, mouthwash, an appetite stimulant (iv), and as a flavouring for ginger beer.

ROSACEAE

Chrysobalanus icaco L.

Local names: Fat Pork, 'Z'Icaque' (Dominica); Coco Plum (Jamaica); 'Icaque', 'Pomme Z'Icaque' (Fr. West Indies).

Found in the West Indies, Central America and South America (i). In Dominica it frequently grows in coastal areas, usually in red clay soils, in littoral vegetation.

    A bushy evergreen shrub about 3–4 m high with shiny ovate leaves. The white flowers are in cymes. The fruit is a drupe and turns white, pink, or dark purple when mature. It is edible and faintly rose-scented.

    The wood was used by the Caribs to make torches (xix).

Medicinal uses The plant was used for dyspepsia and diarrhoea, after boiling the fruit and leaves (iv).

Rubus rosifolius Sm.*

Local names: Wild Strawberry; 'Fraise' (Dominica); 'Framboisier' (Fr. West Indies).

Native of S.E. Asia and found in parts of the West Indies (i). In Dominica, it is found at the middle to upper elevations, on open grasslands, cultivated pastures and by roadsides.

    This shrub is about 1.5 m in height with a spreading habit. The flaccid leaves are light green, compound and have serrated margins. Both leaves and stems are covered with prickles. The five-petalled flowers are white. The fruit turns scarlet when mature.

    The fruit is edible and can be eaten raw, or cooked and used for jam or jelly.

Medicinal uses Used in baths against 'piai', or 'l'amarage'.

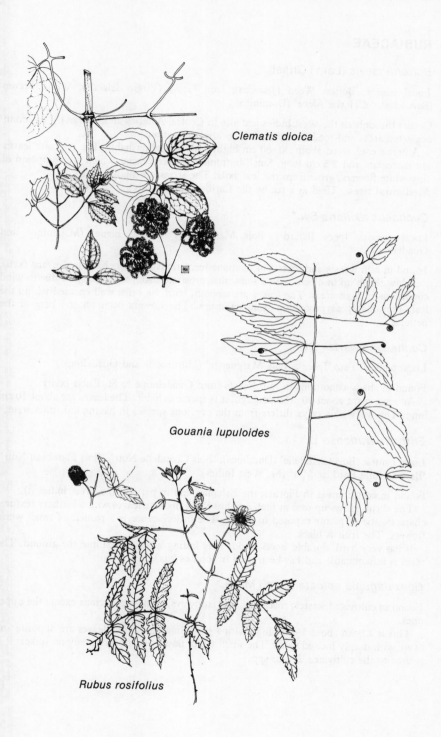

Clematis dioica

Gouania lupuloides

Rubus rosifolius

RUBIACEAE

Borreria laevis (Lam.) Griseb.*

Local names: Button Weed (Jamaica); Iron Grass (Virgin Islands); White Broom (Barbados); 'Z'Herbe Akwe' (Dominica).

Occurs throughout the West Indies and also in Central America and Hawaii (i). Found in neglected areas and roadsides at middle to lower elevations.

A herbaceous weed about 30–60 cm high, with straggling habit. The opposite leaves are lanceolate and 2.5 cm long. Small buttonlike inflorescences, which are composed of tiny white flowers, grow from the leaf axils. The stems are slightly hairy.

Medicinal uses Used as a tea by the Caribs for colds (xix).

Cephaelis axillaris Sw.*

Local names: 'Ipéca Bâtard', 'Bois Marguerite', 'Graine Bleue' (Martinique and Guadeloupe).

Found in rain forests and forests of montane formations from St. Kitts to Guyana (xvii).

A low shrub up to 45 cm with a somewhat prostrate habit, which may be due to wind conditions in high areas. The leaves are smooth, with the veins well encased within the leaf, presumably as protection from exposure. The flowers occur at the base of the petioles.

Cephaelis swartzii DC*

Local names: 'Faux Ipeca', 'Bois Marguerite' (Martinique and Guadeloupe).

Found in shady conditions in rain forests from Guadeloupe to St. Lucia (xvii).

An erect plant about 90–150 cm high, it is sparse in habit. The leaves are about 10 cm long. The waxy blue calyx differs from the previous species in having a distinct stem.

Erithalis fruticosa L.*

Local names: 'Bois Chandelle' (Dominica); 'Bois Chandelle Noir', 'Bois Flambeau Noir', 'Bois d'Huile Bord de Mer' (Fr. West Indies).

Found in coastal areas in Florida, the Bahamas and in parts of the West Indies (i).

The shrub grows up to 5 m high and has shiny obovate leaves with a leathery texture, characteristic of plants exposed to winds. The inflorescence is a panicle of small white flowers. The fruit is black.

It has very hard, durable wood, and is long lasting when driven into the ground. The resin is inflammable and has been used in torches (xix).

Gonzalagunia spicata (Lam.) Maza.*

Found in cultivated wastes, roadsides, and pathways, at most elevations except the upper ones.

This is a bush about 1–2 m high with a straggling habit. The leaves are opposite and ovate with deeply incised veins. The small white flowers occur terminally on spikes. It is related to the cultivated *Pentas* spp.

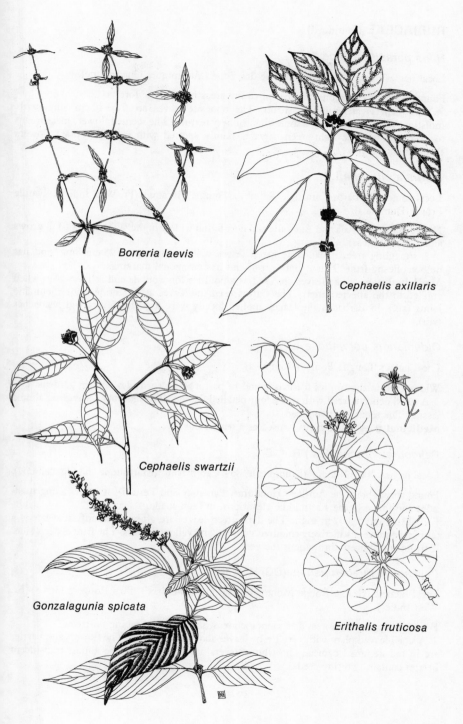

Borreria laevis

Cephaelis axillaris

Cephaelis swartzii

Gonzalagunia spicata

Erithalis fruticosa

RUBIACEAE (continued)

Hillia parasitica Jacq.*

Local names: 'Jasmine Bois', 'Jasmine des Bois' (Martinique, Guadeloupe).

Found at upper elevations on trees in sunny areas in montane forests.

This plant is an epiphytic shrub. The opposite leaves are 7.5–10 cm long, with indistinct venation and with a somewhat leathery texture. The scented flowers measure up to 10 cm wide. The light-green calyx is faintly spotted with red. The plant flowers throughout the year.

Morinda citrifolia L.*

Local names: 'Rubarbe Caraïbe', 'Bilimbi', 'Pomme Macaque' (Fr. West Indies); 'Feuille Froid' (Dominica).

Native of tropical Asia and Australia and now found widely in the West Indies (i). It grows along coastal areas.

A straggling tree, it has large green leaves which grow up to 25 cm long, and has bulbous fleshy fruit. This is light green and has hexagonal markings.

Medicinal uses The leaves are used as a poultice for wounds and relieves pain when applied to the affected areas. The Caribs used the leaves as a treatment for rheumatic joints (xix). In Barbados the leaves applied locally were used for fevers and headaches (viii).

Oldenlandia corymbosa L.*

Local name: 'Langue Poule' (Dominica).

Widespread in the tropics (i). It is found in pastures, by roadsides, and in gardens.

A small delicate weed with white four-petalled flowers. The lanceolate leaves are almost sessile. The seeds are two-lobed.

Medicinal uses The plant is used as a tea for colds.

Palicourea crocea (Sw.) R. & S.*

Local names: 'Bois l'Encre' (Dominica); 'Bois Cabrit Noir' (Martinique and Guadeloupe).

Found throughout the Antilles, Honduras, Paraguay and Peru (i). It grows along roadsides and forest borders at middle elevations, in semi-shade.

A shrub about 1–2 m high. The dark-green leaves are glossy. The inflorescence is a terminal panicle distinctively coloured with red and yellow flowers. The fruit is deep blue. It fruits and flowers throughout the year.

Psychotria guadelupensis (DC) Howard*

Local names: 'Graine Rouge Montagne', 'Graine à Perdrix', 'Bois Rouge à Grive' (Fr. West Indies).

Found in upper rain forest and montane areas, sometimes climbing on trees.

A small-leaved plant with very fleshy leaves about 4.5 cm long. The flowers and berries are in red-stemmed cymes. The white flowers are 6 mm long. The almost translucent berries contain two tiny seeds.

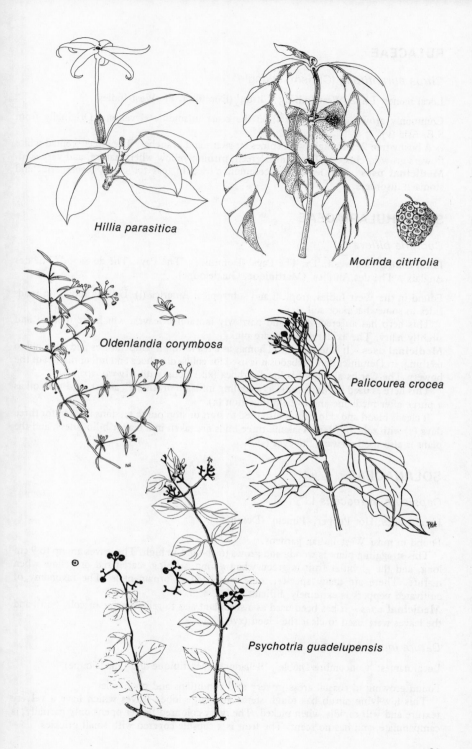

Hillia parasitica

Morinda citrifolia

Oldenlandia corymbosa

Palicourea crocea

Psychotria guadelupensis

83

RUTACEAE

Citrus aurantifolia (Chrism) Swingle*

Local names: Lime (West Indies); 'Citron' (Dominica, Fr. West Indies).

Commonly grown commercially from the coast to middle elevations. Originally from S.E. Asia (v).

A bushy tree 2–5 m high, with spines on stems and branches. The sweet-scented white flowers grow in clusters. The fruit is green, turning yellow when mature, and very acid.
Medicinal uses The leaves are commonly used in teas for colds, fevers, 'flu, and stomach disorders.

SCROPHULARIACEAE

Capraria biflora L.*

Local names: Goatweed; 'Du Thé Pays' (Dominica); 'Thé Pays, Thé de Sante' 'Thé des Anglais', 'Thé des Antilles' (Martinique, Guadeloupe).

Found in the West Indies, tropical and subtropical America (i). Usually found on road-sides in somewhat poor soil.

This herb has an erect habit and narrowly lanceolate leaves, which are toothed and slightly hairy. The axillary flowers are pink.
Medicinal uses It is used in St. Thomas as a tea for high temperatures and as an aid to teething. In Dominica and Barbados it is used for colds, diarrhoea and inflammation of the bowels. The Caribs also used it for diarrhoea (xix) along with *Myrcia citrifolia*.

This herb is used in Martinique as a cooling infusion or 'tisane', and is used with oil as a purge after medicine and to help digestion (x).

It cleans blood and skin, and is also used to start or stop purge sessions if taken for three days (x) with salt. In Marie Galante three buds are taken in a tea for biliousness, and the plant is also used to wash the eyes (xvi).

SOLANACEAE

Capsicum frutescens L.*

Local names: Hot Pepper; 'Piment' (Dominica).

Found in most West Indian gardens.

This straggling plant is woody and grows to about 1 m high. The leaves are up to 9 cm long, and the globular fruit is green when young, turning scarlet red or yellow when mature. There are many species, some cultivated as ornamentals. The taxonomy of cultivated peppers is extremely difficult (i).
Medicinal uses It has been used as a stimulant and to relieve gripe or colic (viii), and the leaves were used to clear the blood (xvi).

Datura innoxia Mill.*

Local names: 'Concombre Diable'; 'Belladone' (Martinique and Guadeloupe).

Found growing in coastal areas in very arid conditions and poor soils.

This low-lying shrub has tough stems and large lobed leaves which have a velvety texture and wilt rapidly when picked. The ivory-coloured flower opens only partially, is campanulate, and has no scent. The fruit is a capsule covered with small prickles.

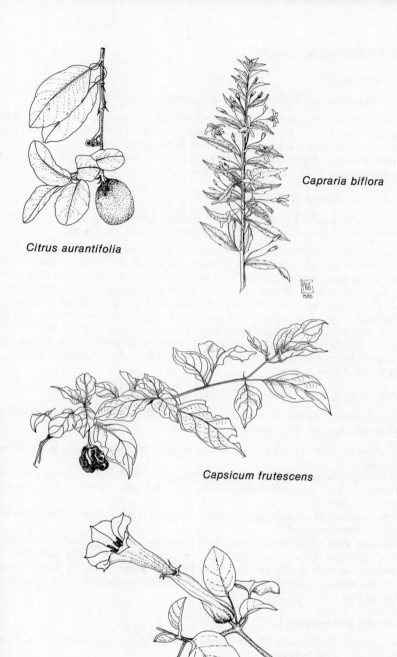

Citrus aurantifolia

Capraria biflora

Capsicum frutescens

Datura innoxia

SOLANACEAE (continued)

Datura metel L.*

Local name: Prickly Bur (St. Thomas, St. Croix).

An annual about 1.5 m high with smooth stems. The leaf margins are wavy or irregular. The calyx is 6–7 cm and has five teeth. The corolla is ivory-coloured, sometimes mauve-tinted, about 10–12 cm long. The ovoid capsule has burrs.

Medicinal uses A **poisonous** plant, particularly the seeds. The leaves, when bruised, are applied externally for ulcers or for arthiritis.

Datura suaveolens H.B. ex Willd.*

Local names: Angel's Trumpet (Jamaica, Trinidad and Tobago); 'Fleur Trompette', 'Trompette du Jugement' (Martinique and Guadeloupe); Jimson Weed.

A native of Brazil (i), found at middle elevations growing on roadsides.

It is a small tree with broad alternate leaves and slightly fleshy stems. The flowers are about 20 cm long, with slightly reflexed petals. They are distinctive, not only for their very pervading sweet scent, but also because of the slight change in colour as they mature. They are ivory white at first, then turn apricot-coloured as they begin to shrivel.

Medicinal uses The leaves and flowers of this tree are apparently **narcotic**. Other species were used by Indians in South America as hallucinatory agents — possibly *Datura stramonium*, which is used in Marie Galante by smoking the dry leaves, for asthma, or else passing the leaves over fire and placing them on the forehead to counteract neuralgia. It is **dangerous** if used to excess.

Solanum americanum Mill. var. nodiflorum (Jacq.) Edwards*

Local names: 'Z'Herbe Amère' (Dominica, Fr. West Indies); 'Herbe à Calalou', 'Agouman', 'Agouma' (Fr. West Indies).

Cosmopolitan distribution (i), this plant grows along roadsides and in gardens at coastal elevations.

A straggling herbaceous weed with simple leaves about 10 cm long, sometimes notched at the base. There are small clusters of tiny white flowers. The green or red berries are shiny and round.

It may be used as a substitute for spinach, is a good rabbit fodder, and was used by the Caribs as a potherb (xix).

Medicinal uses Leaves, stems and flowers were used in cooling teas for inflammation caused by fatigue and were taken for nine days. Also used in refreshing baths (xvi). Used in cooling teas with *Stachytarpheta jamaicensis* and *Leonotis nepetifolia* for inflammation (x), in Martinique. It is also used as a purge with *P. oleracea* and *Peperomia pellucida* when boiled in beer (x).

Solanum bahamense L.*

Local names: Cankerberry (Virgin Islands); 'Bois Teurtre' (St. Barthelemy); 'Picanier Femelle' (Fr. West Indies).

Found in sunny coastal areas, this bushy shrub grows about 1–2 m high. The leaves are 8–10 cm long, narrow and smooth. The white flowers are about 2.5 cm in diameter, with bright yellow stamens. The berries are bright scarlet.

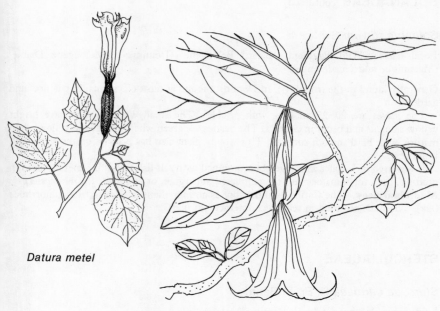

Datura metel

Datura suaveolens

Solanum americanum

Solanum bahamense

SOLANACEAE (continued)

Solanum torvum Sw.*

Local names: Wild Eggplant, 'Bâtard Belongene' (Dominica); 'Melongene Diable' (Martinique and Guadeloupe).

Generally found in the tropics (i). In Dominica found at most elevations, in pastures and gardens.

The leaves are 20–25 cm long with prickles. The small white flowers have bright yellow stamens and occur in clusters. The berries are green when young, and yellow when mature, with hard smooth surfaces. This woody plant also has prickles on the stems and branches.

It was used in Jamaica as a vegetable combined with salt fish (iv). It is generally used in grafting, both for tomatoes and the cultivated aubergine, or eggplant. (*S. melongena*).

Medicinal uses The Caribs used the root as an ingredient in a tea to treat gonorrhoea (xix). The buds are used in teas to counteract 'flu (xvi).

STERCULIACEAE

Sterculia caribaea R. Br. in Benn.

Local name: 'Mahot Cochon' (Dominica).

Found at middle to upper elevations, in forests.

A medium-sized to tall tree, erect, with a distinctive leaf. This has different stages of development. The young leaves may be lobed. The middle-aged leaves have three distinct lobes, whilst the mature leaves are entire with no lobes. The yellow inflorescence has a five-parted calyx (xvii). The hard fruit is a gourd-shaped capsule. The exterior of the fruit splits open to show four or five shiny brown seeds, shaped like olives.

The wood was used for staves and boards, and is now only used for interior work because it is soft and susceptible to water.

Waltheria indica L.*
(syn. *Waltheria americana* L.)

Local names: Buff Coat (Virgin Islands); 'Guimauve' (Martinique, Guadeloupe).

Widespread in the tropics and subtropics (i). Found in Dominica in low grassy areas.

A herb with a stiff hairy stem. The hairy leaves are about 7.5 cm long. The flowers are clustered closely around the stem both in axillary and terminal positions.

Medicinal uses The plant was used as a cooling drink and by boiling the leaves and stems as a hot tea for colds. It was also used for cystitis.

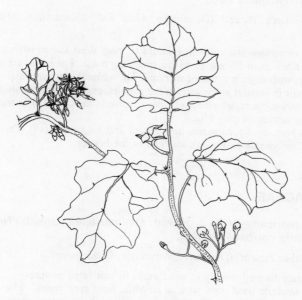

Solanum torvum

Waltheria indica

SYMPLOCACEAE

Symplocos martinicensis Jacq.*

Local names: 'Graines Bleues' (Dominica); 'Caca Rat' (Martinique, Guadeloupe, St. Lucia).

Found at middle elevations and cultivated wastes in most West Indian islands.

A tree which grows about 15 m high with a spreading habit. The leaves are very shiny on the upper side, with slightly crenated margins. The white flowers are small and sweet scented. They occur in clusters at the axils of the petioles, and have five petals with many small stamens adnate to them, which appear like a crown in the centre. The small oblong berries are green, turning to navy blue.

The wood has been used for interior house work and for boards (xx). Trembleurs, a type of thrush (*Cinclocerthia ruficauda*), are fond of the berries.

THYMELAEACEAE

Daphnopsis americana (Mill.) J. Johnst. spp. *caribaea* (Griseb.) Nevling (syn. *Daphnopsis caribaea* Griseb.)

Local name: 'Mahot Piment' (Dominica, Martinique, Guadeloupe).

Found in secondary forests, usually in windbreaks in cultivated pastures.

A spreading medium-sized tree with a smooth, light-grey trunk. The leaves are alternate, lanceolate and entire. They are smooth and light green, therefore readily identifiable from the normally darker-green foliage around. The mass of bordering seedlings might lead one to suspect this is only a small bush, but the trunk can grow to 45 cm in diameter. The small white flowers are in clusters at the end of the branches, and the fruit is a small, oblong, white berry. The twigs have a habit of branching out and subdividing so that most of the leaves appear in whorls at the ends, giving the tree a loose, graceful appearance.

The bark is fibrous and extremely tough and is often used to make ropes or 'colliers' for small livestock such as sheep or goats.

Medicinal uses Teas have been made from the leaves for baths against 'piai' or evil spells. They were often used as an ingredient in baths for other rituals. The bark was sometimes steeped with *Inga laurina* and given as a tea to encourage lactation in women (xix).

TILIACEAE

Sloanea berteriana Choisy

Local name: 'Châtaignier Petit Coco' (Guadeloupe).

The Châtaigniers are imposing rain forest trees found in some of the Lesser Antilles and in Puerto Rico, Hispaniola and tropical South America.

This tree is evergreen, about 15–18 m high, with a cylindrical trunk above its buttresses. The leaves are oblong, about 11–25 cm long and 9 cm broad. The inflorescence is a raceme, and the fruit is a woody oval capsule, with hairs.

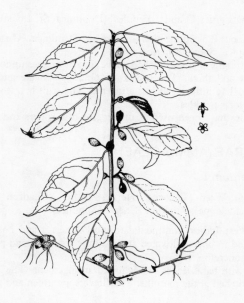

Symplocos martinicensis

TILIACEAE (continued)

Sloanea caribaea Kr. Urb.

Local name: 'Châtaignier Petite Feuille' (Guadeloupe, Dominica, St. Lucia).

Found in Dominica, Guadeloupe, St. Lucia, St. Vincent and Grenada.

The trunk is straight or slightly lobed above the buttresses, which are as high as 4 m. The tree itself grows up to 30 m high. The crown is dense and the leaves oblong, elliptical, 6–11 cm long and 2.5–5 cm broad. The inflorescence is either terminal or axillary and is a cyme. The fruit is a smooth woody capsule.

Sloanea dentata

Local name: 'Châtaignier Grandes Feuilles' (Dominica, St. Lucia).

An evergreen about 25–30 m high, found in Dominica, Guadeloupe, Martinique, St. Kitts and tropical South America.

The trunk is straight or slightly lobed above thin buttresses up to 2.5 m high. The bark is reddish-brown and thick. The leaves are oblong, obovate and 21–30 cm long and 14–18 cm broad. The inflorescence is a raceme. The fruit is a woody capsule with four valves and with long prickles.

All three species have been used as timber, for mill rollers and house interiors. (xx).

UMBELLIFERAE (AMMIACEAE)

Eryngium foetidum L.*

Local names: Fit Weed (Tobago); 'Herbe Puante', 'Chadron Béni', 'Herbe à Fer' (Martinique, Guadeloupe); 'Chadon Benée' (Dominica).

Found in the West Indies, continental tropical America and now in West Africa and Uganda (i). Found in once-cultivated areas, sunny locations and particularly growing in old rubble and concrete.

A small herb with basal leaves in a whorled arrangement. The leaves are long, almost spatulate, and notched at the margins. The flowers are green and grow in small terminal clusters.

Medicinal uses The plant is used for fevers and chills. The Caribs used the herb as a cure-all (xix), while Jamaicans used it for colds and fits in children. The plant was also rubbed on the body for fainting and convulsions (viii), and used in cases of vomiting, diarrhoea and fever. Also used after purges (x).

URTICACEAE

Laportea aestuans (L.) Chew.*

Local names: Stinging Nettle, 'Ortie Brulante' (Guadeloupe, Martinique, St. Maarten, St. Barthelemy); 'Zootie' (Dominica).

General in the tropics and found in shady and semi-shade conditions, frequently in vegetable gardens from the coast to lower-middle elevations.

This weed is slightly reddish in the stem, and has opposite heart-shaped leaves with serrated margins and stinging hairs on both the leaves and stems.

Medicinal uses The leaves were mixed with oats and used for 'stricture' in Barbados (viii). This sometimes caused difficulty in urination.

Eryngium foetidum

Laportea aestuans

URTICACEAE (continued)

Pilea microphylla (L.) Liebm.

Local names: Lace Plant (Jamaica); 'Teigne', 'Ti-Teigne', 'Ti-Teigne Blanc' (Guadeloupe, Martinique, St. Barthelemy, St. Maarten).

Found generally in the West Indies, and in Dominica, in gardens, rubble, stone steps, or potted plants.

A low straggling herb with small (12 mm) heart-shaped leaves which are light green and somewhat membranous. The flowers are tiny and pinkish or greenish, and the stem is pinkish.

Medicinal uses It has been used in Jamaica (iv) for asthma, in the Grenadines (iv) for diarrhoea, and in Barbados (viii) for inflamed bowels.

VERBENACEAE

Citharexylum spinosum L.*

Local names: 'Coquelet', 'Bois Côtelette' (Dominica); 'Bois Carré', 'Bois de Fer Blanc', 'Bois Cotelette' (Martinique, Guadeloupe).

A medium-sized tree, found in secondary growth.

This tree has leaves with reddish petioles. Its small white flowers are in terminal spikes. The berries are green when young, turning black on maturity.

The wood is very hard and is used in house building for rafters and similar support structures.

Medicinal uses The plant is used in baths for fatigue. A similar species, *C. fruticosum* is also used in refreshing baths, or the leaves, when dipped in castor oil, may be used as a poultice. (xvi).

Clerodendrum aculeatum (L.) Schlecht*

Local names: 'Amourette' (Guadeloupe); 'Thé Bord de Mer' (Guadeloupe, Martinique, St. Maarten, St. Barthelemy).

Found in dry coastal conditions in full sun and poor soil.

A small evergreen shrub with glossy opposite leaves. The flowers are in a terminal cluster, with five joined petals and protruding stamens and style. The flowers are white, the stamens and style dark pink. The flowers have a noticeable odour. Each calyx is small, slightly reflexed, and five-parted.

Clerodendrum philippinum Shauer*

Local names: 'Rosalba' (Dominica); 'Herbe Puante' (Fr. West Indies).

A native of China, this plant is now common in the tropics (i). It grows in moist lowlands and at middle elevations, and is common along roadsides.

A shrub with a spreading habit, it has large opposite leaves on thick stems which are covered with fine hairs. The pink flowers are in terminal cymes and cupped by a large calyx. They have a faintly sweet scent.

Medicinal uses The flowers were used by the Caribs to cure headaches by inhalation (xix).

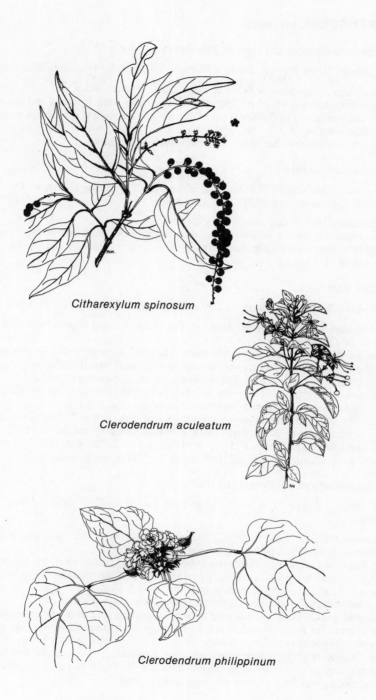

Citharexylum spinosum

Clerodendrum aculeatum

Clerodendrum philippinum

VERBENACEAE (continued)

Clerodendrum speciosissimum Van Geert ex Morren*

Local names: 'Herbe Puante', Herbe à Madame Villaret (Martinique, Guadeloupe).

Native of Java (i) this plant is found in many areas along roadsides.

A spreading shrub with 30 cm long ovate leaves. The stem is flaccid and the whole plant is soft textured. The flowers occur terminally in panicles, and the calyxes are purple with scarlet corollas.

It is a cultivated plant but seems to be an escape in Dominica.

Cornutia pyramidata L.*

Local names: 'Bois Caral', 'Bois Cac', 'Bois Savane', 'Bois Cassave', 'Mouri Douboute' (Martinique, Guadeloupe); 'Bois Carré' (Dominica).

A medium-sized tree found in secondary forests.

The distinctive feature of this tree is that all the terminal branches and stems are noticeably square, becoming rounded toward the trunk (xi). The light-blue flowers occur at the end of the branches.

Lantana camara L.*

Local names: Wild Sage, 'Ma Bizou', 'Mal Visou'; Mille Fleurs (Marie Galante).

Found in tropical and subtropical America. The shrub is found from coastal areas up to 80 m, usually along the roads.

This fairly bushy shrub grows 1 m or more in height, but there are dwarf species which cover the ground. The leaves are opposite, ovate and rough textured. The yellow and red flowers grow in terminal cymes. The stems are woody and quadrangular, and the leaves are aromatic when crushed. The fruits are small berries and are either blue or black when mature. It has been used as an ornamental in some gardens.

(*L. crocea* Jacq. has flowers which are yellow and orange, and *L. involucrata* L. has mauve flowers with yellow throats. *L. reticulata* Pers. has plain mauve flowers.)

Medicinal uses The flowers and buds are taken in a tea for 'flu and chills, and are used in baths; *L. involucrata* flowers are used in teas for high blood pressure (xvi).

Stachytarpheta jamaicensis (L.) Vahl.*

Local names: 'Verveine', 'Verveine Queue de Rat' (Fr. West Indies); 'Veng Veng' (Dominica).

Common in the new world tropics (i), found in most areas in Dominica where there is cultivation, on roadsides, paths and at most elevations.

An annual or perennial, with an erect habit. It has a semi-woody stem. It is about 50 cm high and the dark green leaves are heavily veined. The brilliant blue flowers are on spikes. There are usually four or five open at one time in each spike.

Bees are attracted to the flowers for nectar, at about noon (xxi).

Medicinal uses The plant was used in St. Thomas and Dominica as a constituent for teas and infusions for colds and fevers. In Barbados it was used as a vermifuge and the leaves were applied to sores and wounds (viii). The stem is used as a vermifuge in teas in Marie Galante (xvi). In Africa it was used for gonorrhoea, eye troubles and sores in children's ears, as well as for heart trouble (iv). The 'Verveine Blanc' (*S. cayennensis*) is used in cooling teas for a women who is nursing (x, xvi).

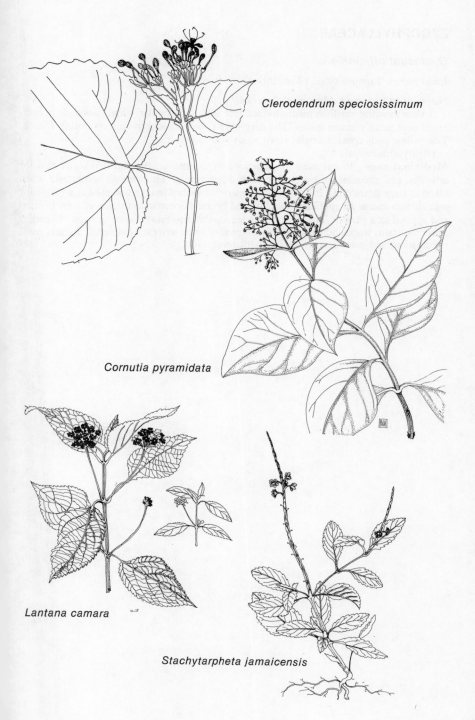

Clerodendrum speciosissimum

Cornutia pyramidata

Lantana camara

Stachytarpheta jamaicensis

ZYGOPHYLLACEAE

Guaiacum officinale L.

Local name: 'Lignum vitae' (West Indies).

Native of tropical and subtropical America.

A slow growing small to medium-sized tree with a particularly hard wood. It is ever-green with small pinnate leaves. The deep-blue flowers bloom in clusters during March. The yellow pods contain small brown seeds with a red aril.

Highly ornamental.

Medicinal uses It was used for abortions by boiling up the leaves (v), and also for arthritis, and rheumatism by boiling with other plants. This caused haematuria when taken in large doses (viii). In Jamaica, the resin was soaked in rum and used as a gargle for sore throats and as an application to cuts and bruises. Sometimes the leaves were bruised and applied as a plaster. The juice of the leaves was also taken for biliousness. In earlier times the resin, wood and bark were considered of value in treating venereal disease, gout, rheumatism and sometimes in intermittent fevers. (iv).

Part 2

MONOCOTYLEDONS

AMARYLLIDACEAE

Hippeastrum puniceum (Lam.) Kuntze

Local names: Easter Lily (Dominica); 'Lis Rouge', 'Fleur Trompette' (Martinique, Guadeloupe).

This plant is native in the West Indies, Central America, and is now found in most tropical countries at elevations of 0–500 m.

A herbaceous plant which grows in a clump. The light-green leaves are about 30 cm long. The showy flowers are about 13 cm across and are bright vermillion with yellow-green centres. The six petals overlap at the base of the flower, and the stamens and style curve upwards. There are commonly two or three flowers on each stem. It flowers February to May. It rarely sets fruit. This plant is cultivated as an ornamental.

Medicinal uses The bulb was used by the Caribs with *Eryngium foetidum* on swellings and sores (viii), and also as an arrow poison (viii). The effects if taken **internally** are gastro-enteritis, nervous disturbance, and dryness in the mouth (viii).

ARACEAE

Anthurium dominicense Schott.*

Found on upper mountain slopes and summits.

A terrestial plant. The leaves are leathery and rather elongated and measure 20 cm long and about 9–15 cm broad. The stems may be up to 35–40 cm long. The stem bears a dark brown, stubby spadix about 4–5 cm long, subtended by a light-green spathe about 4 cm long.

Anthurium grandifolium (Jacq.) Kunth.*

Local names: 'Grande Siguine', 'Siguine' (Martinique, Guadeloupe).

Found in rain forests and secondary forests in Dominica.

It may be either epiphytic or terrestial. The roots are spreading in habit. The heart-shaped leaves are about 70 cm long. The stem bears a thin green spathe and a long dark brown spadix, which measures up to 50 cm long.

Anthurium hookeri Kunth.

Local name: 'Siguine Rouge' (Guadeloupe).

Found in rain forests and montane forests in Guyana and the West Indies (i).

An epiphyte which has large obovate leaves about 30–60 cm long. These grow in a rosette and the plant may measure up to 90 cm across. It has a ribbon-like spathe and a thin, deep-blue spadix.

Anthurium palmatum (L.) Kunth.*

Local names: 'Bénéfice', 'Gagne-Petit' (Guadeloupe and Martinique); 'Mibi' (Dominica).

An epiphytic plant found in shady places at higher elevations and in secondary forests.

A large fleshy vine, it has glossy eight-lobed leaves. The stem has a narrow, greenish spathe and a thick, dark brown spadix measuring up to 20 cm long.

The Caribs used the wood for baskets, scraping it and dying it yellow or purple.

Medicinal uses The bark was used together with scrapings from conch shells as a poultice for yaws (xix).

Anthurium dominicense

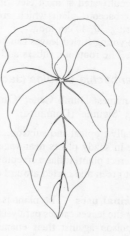

Anthurium grandifolium

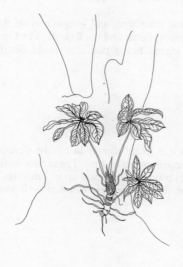

Anthurium palmatum

ARACEAE (continued)

Caladium bicolor (Ait.) Vent.

Local names: 'Dachine Marbré', 'Hâle Bois', Caladium (Dominica); 'Madère Bâtard', 'Petit Madere', 'Calalou Sauvage', 'Chevalier Rouge', 'Palette du Peintre' (Guadeloupe and Martinique).

A native of tropical South America, it is common as an ornamental.

A herbaceous plant with flaccid stems and heart-shaped, single leaves. These are very often distinctively patterned with red or white markings.

Medicinal uses The Caribs used the heated rhizomes as a poultice for extracting splinters (xix).

Colocasia esculenta (L.) Schott.

Local names: 'Dachine', 'Madère', 'Chou de Chine' (Guadeloupe and Martinique); 'Dachine' (Dominica).

A native of India and the Orient, it is now commonly grown in the West Indies, and is widely cultivated at most elevations. There are many varieties.

A herbaceous plant which grows in clumps, the large soft leaves are heart-shaped, measuring 30–40 cm long.

The young furled leaves can be used to make Calalou, a soup widely used in the West Indies. The roots are tubers and are a staple in West Indian cooking.

Dieffenbachia seguine (Jacq.) Schott.*

Local names: Dumb Cane, 'Canne Seguine', 'Canne d'Eau', 'Seguine d'Eau', (Guadeloupe); 'Siguine' (Dominica).

Generally tropical and subtropical, the plant is used as an ornamental, and is found as an escape in shady places near water or in thickets, in middle elevations or coastal areas.

An erect plant, it has a creeping fleshy stem with elliptical variegated leaves. The spathe is light green and curled around a pale-yellow spadix. The plant exudes a milky sap when cut.

Medicinal uses The plant is **poisonous** if chewed, and livestock avoid it normally. Eating the leaves causes paralysis of the mouth, throat and vocal cords. The Caribs used it as a poison against their enemies, and legend has it that they could destroy several generations with one dose (xix).

BROMELIACEAE

Aechmea smithiorum Mez in DC.*

Endemic to Lesser Antilles. In Dominica it is found usually in middle elevations.

A plant with a rosette of long light-green leaves (up to 1 m). These have finely serrated margins. The inflorescence is a cylindrical spike, on a peduncle up to 1 m long. The many pink and lilac bracts which surround the flowers are very distinctive. It blooms from September to November.

Dieffenbachia seguine

Aechmea smithiorum

BROMELIACEAE (continued)

Glomeropitcairnia pendulifera (Grizeb.) Mez.*

Local name: 'Ananas Grand Bois' (Fr. West Indies).

Found at upper elevations, it is the largest of the bromeliads.

The yellowish inflorescence grows from the stout rosette of leaves, which is about 2 m in diameter, and up to 3 m in height.

Guzmania dussii (Baker) Mez.*

Local name: 'Ananas Grand Bois' (Fr. West Indies).

This bromeliad is not common, found only in upper montane forest and on mountain tops.

An epiphytic plant about 60–80 cm broad. The leaves are light-green, leathery and form a rosette which measures about 1.5 m across. The inflorescence stands 65 cm high and has a compact habit. The bracts point upwards and are heavily streaked with pink, shading towards green. This colouring becomes more pronounced toward the top. The flowers are buttercup-yellow. It blooms in March.

Guzmania plumieri (Griseb.) Mez.*

Local name: 'Ananas Sauvage Montagne' (Fr. West Indies).

Found on mountain slopes in Dominica.

This bromeliad has a rosette of leaves which turns maroon at the base. The plant is about 1 m across, and the inflorescence is 60 cm or more tall. This has a green stem and dark red bracts which contain the yellow flowers. This bright colouring is more pronounced in the plant the further up the slopes one finds it — due, presumably, to increased sunlight. Lower down the slopes the inflorescence tends to stay greenish, with only slightly maroon coloured bracts, and is not as spectacular. It blooms in March.

Pitcairnia angustifolia Sol.*
(syn. P. latifolia Sol., syn. P. gracilis Mez.)

Local names: 'Bâtard la Pite' (Dominica); 'Ananas Rouge Bâtard' (Martinique and Guadeloupe).

Native of South America and found in Dominica and other West Indian islands. It is fairly common along the roads.

A terrestial plant with long narrow leaves, which are finely hairy on the under-surface. The inflorescence is a brilliant scarlet, on panicles. Each bract is two-lipped. The plant flowers from March to September.

Medicinal uses The small hairs on the underside of the leaves were scraped and placed on a child's navel after the umbilical cord was cut (xix).

Pitcairnia spicata (Lamarck) Mez. var. sulphurea (Andrews) Mez.

Found on mountain slopes.

This bromeliad has the same habit as those found commonly in lower elevations; the leaves are similarly linear. They are dark green and distinguished from the P. angustifolia by their flaccid and more spreading habit. The inflorescence is also scarlet.

Glomeropitcairnia pendulifera

Guzmania dussii

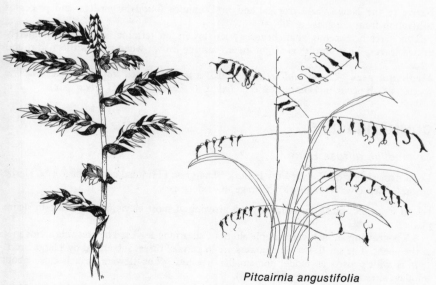

Guzmania plumieri

Pitcairnia angustifolia

105

BROMELIACEAE (continued)

Tillandsia usenoides L.

Local names: Old Man's Beard (Trinidad and Tobago); Spanish Moss (Virgin Islands); 'Barbe à l'Arbre' (Guadeloupe).

Found growing on trees from Florida to as far south as Chile.

An epiphyte with hanging mat-like habit spreading over trees in lacy fronds. The greyish-silver leaves are very slender. It has tiny green flowers.

Tillandsia utriculata L.

Local names: Wild Pine; 'Ananas Sauvage' (Guadeloupe and Martinique).

A common bromeliad in Dominica that is found from coastal to lower middle elevations particularly in dry conditions. It covers the branches of trees.

Epiphytic, and varying in size from 6 cm to 1 m long, the light greyish-green leaves are numerous, linear and acuminate, with smooth margins. They are covered with tiny scales, and the lower leaves tend to be reflexed. The yellow-brown inflorescence is in a spike.

CANNACEAE

Canna indica L.*
(syn. C. coccinea Mill.)

Local names: Canna Lily; 'Calenda', 'Balisier Rouge', 'Balisier à Chapelets' (Martinique and Guadeloupe).

Native to the New World tropics, and in Dominica found in gardens and places of cultivation from the coast to 300 m.

This erect herbaceous plant bears scarlet flowers on terminal stalks with pinnately veined leaves. The capsules are light green, turning brown when ripe, with small round seeds.

Medicinal uses *Canna edulis*, called 'tous les mois', is used for stomach-aches. The species which occur in Dominica are *C. indica*, *C. lambertii* and *C. edulis* (xix).

COMMELINACEAE

Commelina diffusa Burm f.*

Local names: French Weed (West Indies); Watergrass (Trinidad and Tobago); 'Z'Herbe Grasse', 'Zeb Gwa' (Dominica); 'Curage' (Guadeloupe).

Found in tropics and subtropics (i), and growing at most elevations, thriving in damp places.

A low-growing annual, with simple alternate sheathing leaves. The creeping stems root at the nodes. The small, sky-blue flowers are in parts of three, subtended by a large bract.

It is widely used as fodder for smaller livestock. The flowers, which close about midday, attract bees for the pollen (xxi).

Medicinal uses It is used as a cooling infusion, and is also an ingredient in protective baths. It is used in a tea for 'flu (xvi).

Canna indica

Commelina diffusa

COSTACEAE

Costus scaber Ruiz & Pav.*
(syn. *Costus cylindricus* Jacq.)

Found in Dominica in forested areas from middle to upper elevations.

A tall straggling plant (up to 3 m) with alternate, narrowly elliptic leaves and a cylindric inflorescence, which is made up of conspicuous red bracts from which a yellow flower emerges. The stems are slightly twisted.

Costus speciosus (Koenig.) J. E. Smith.*

Found in Dominica along roadsides and in cultivated wastes at upper elevations.

This medium-sized plant (about 2 m) has alternate, narrowly elliptic leaves on a reddish stem. The inflorescence is ovoid made up of deep red scale-like bracts. The flowers are white and delicate but have no scent.

Costus sp.*

Local names: 'Poivre Ginet', 'Poivre de Guinée' (Dominica).

A low spreading herbaceous plant with reddish stems which are slender and drooping and up to 1.5 m long. The leaves are 14–15 cm long. The round flowers grow at the base of the plant from stout reddish bracts, and they have a sweet scent. The corolla is about 6 cm in diameter, and is white with a yellow throat.

Medicinal uses   The leaves of the plant are boiled in a tea for gas.

*Curcuma roscoeana**

Found at middle elevations along roads.

This plant, about 30 cm high, has elliptical leaves with an interesting pattern of venation. There is a dark maroon marking on either side of the light-green central rib and the veins are a darker green than the rest of the leaf. These leaves shoot from the ground. The inflorescence is cone-shaped, with whitish cup-like bracts tinged at the upper edges in deep maroon or purple. The light yellow flowers are tubular with a deeper yellow streak at the lower lip. It blooms in May. It is closely related to tumeric (*Curcuma domestica*).

CYCLANTHACEAE

Carludovica insignis Duchass.*

Local name: 'Z'ailes Mouches' (Dominica, Fr. West Indies).

Found from South America to Central America and the Lesser Antilles (x). Found at middle to upper elevations in forested areas.

The terrestrial plant bears palm-like leaves, each of which is lobed into two parts. The venation is strongly parallel, and the leaves may be up to 150 cm long. The inflorescence is a dark brown spadix with the crowded flowers in spirals on its surface. It is 12.5 cm long and about 2.5 cm broad.

The leaves are used as thatching for overnight shelters. They can also be used for lining pots and baskets.

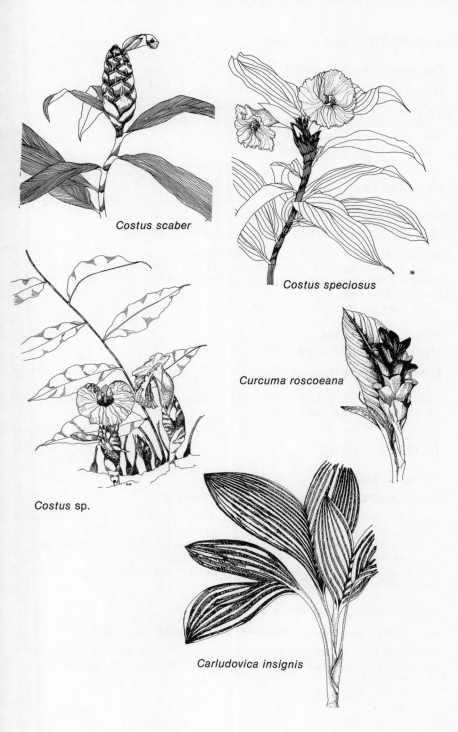

Costus scaber

Costus speciosus

Curcuma roscoeana

Costus sp.

Carludovica insignis

CYCLANTHACEAE (continued)

Carludovica plumieri Kunth.*

Local name: 'Z'ailes Mouches' (Dominica, Fr. West Indies).

Found at middle to upper elevations in forests.
    This species is epiphytic. The roots are fibrous and the leaves, also lobed, are smaller than those of *C. insignis*.
    This plant has the same uses as *C. insignis*.

CYPERACEAE

Rhynchospora nervosa (Vahl) Boek.

Local names: 'Z'herbe Pailasse' (Dominica); Star Grass (Jamaica).

Native to the New World, and found in damp areas.
    A perennial sedge about 15 cm high, with an erect three-sided stem rising to form a rosette of green leaves, which are white at the base. The inflorescence is in spikelets and is attractive to bees (xxi).
    The plant has been used for stuffing mattresses.

Scleria latifolia Sw.

Local names: 'Z'herbe à Couteau'; Razor Grass.

Found in forests, clearings and cultivated wastes at middle to upper elevations. Common in tropical America.
    It is distinctive for its broad, razor-edged leaves, and the stem is distinctly three-sided. The inflorescence is a panicle.

DIOSCOREACEAE

Dioscorea alata L.

Local names: White Yam; 'Igname Blanche' (Dominica).

A native of the Old World tropics and planted widely in the West Indies for its edible starchy root. Approximately nine species of *Dioscorea* appear in the West Indies.
    This vine has green, heart-shaped alternate leaves about 15 cm long and 10 cm broad. The stem is four-sided.
    Although there are many varieties of *Dioscorea*, this one is considered to have a good flavour and consistency.

Dioscorea polygonoides H. & B.*

Local names: 'Yam Wa Wa' (Dominica); 'Igname Sauvage' (Fr. West Indies).

A wild yam which will carry its vine to the treetops. The leaves are shiny, and the stem is reddish-brown. The winged seeds hang in clusters at intervals along the stem.

Carludovica plumieri

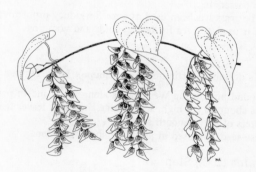

Dioscorea polygonoides

GRAMINAE (POACEAE)

Arthrostylidium excelsum Griseb.

Local name: Bamboo Grass.

At middle to upper elevations, this climbing, bamboo-like grass is found in forests.

The stems are slender but cable-like. It is like the large bamboo (*Bambusa vulgaris*) in appearance and the leaves have the same formation. They form a thicket-like mass.

Bambusa vulgaris Schrad.

Local names: Bamboo, 'Bambou' (Dominica, St. Lucia, Fr. West Indies).

Native of the Old World tropics, found commonly in the New World and at lower to middle elevations.

Growing in clumps along roadsides and forming graceful plumes of feathery leaflets, these plants were used as soil erosion controls on steep roadsides, riverbanks and in gardens. Many are now being cut to clear ground for more gardens, or during road maintenance. The effects of this policy on steep land will cause some soil erosion problems.

Bamboo is widely used in Dominica for basket-work, fish pots, poles, temporary guttering and for fencing.

Cenchrus echinatus L.*

Local name: Burr Grass.

A common annual throughout the West Indies.

The leaves are narrowly lanceolate on the stems. This grass has small brown spiny fruit which sticks easily to passers-by.
Medicinal uses The grass has been used as a tea for kidney problems in St. Thomas, and in Jamaica (iv) it is used as a tea for fevers, colds and vomiting.

Coix lacryma-jobi L.*

Local names: 'Larmes de Job' (Martinique, Guadeloupe); Job's Tears.

A native of Asia (i) found now in the West Indies along roads and in pastures.

A herbaceous grass about 60 cm high. It has large sheathing leaves from which stalks emerge carrying clusters of grey, smooth seeds.

Ornamentally, these seeds are used in jewellery and decoration.

Cymbopogon citratus (D.C.) Stap.

Local names: Lemon Grass; 'Citronnelle' (Dominica, Martinique, Guadeloupe).

Found in waste places, in fields, and along roadsides anywhere from lower to fairly high elevations.

A medium-sized clump of grass about 1 m in diameter, the leaves are thin, blue-green in colour with very slight serrations at the margins and, when picked, emit a strong lemon-like smell. The inflorescences are on long stalks, brownish in colour, and in fairly loose panicles.
Medicinal uses A commonly used plant in hot teas, for colds or chills. The plant has been used in baths against poisoning (x).

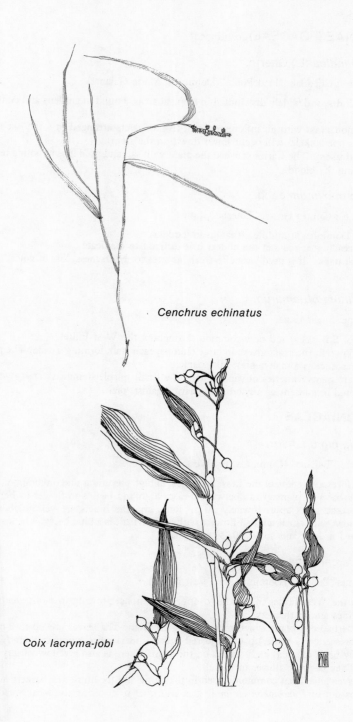

Cenchrus echinatus

Coix lacryma-jobi

GRAMINAE (POACEAE) (continued)

Eleusine indica (L.) Gaertn.*

Local names: Cheddah; 'Pied Poule' (Dominica, Marie Galante).

Native to India, and widely distributed in tropical areas. Found in gardens and cultivated areas.
    A common grass with an inflorescence of five spikelets arranged in a fan-like shape. Sometimes one spikelet will occur lower down on the stem.
Medicinal uses   The Caribs crushed the plant with salt and used it as a cooling tea, and for cleansing the blood.

Panicum maximum Jacq.*

Local names: Guinea Grass; 'Z'herbe Guinée'.

Found in Dominica at middle elevation in gardens.
    Fast spreading, it roots at the nodes. It is difficult to eradicate.
Medicinal uses   It is used very effectively as a tea for hoarseness, loss of voice, or sore throats.

Saccharinum officinarum L.

Local name: Sugar Cane.

A native of S.E. Asia, and now common throughout the West Indies.
    Up to 5 m tall, the grass grows in large clumps, each stalk forming a cane. The plume-like inflorescence is a mauve-grey silky panicle.
    There are many varieties of cane. There is a small purple-stemmed variety which is often planted in gardens as a good luck charm against 'piai'.

HELICONIACEAE

Heliconia bihai L.*

Local name: 'Balisier' (Dominica, Fr. West Indies).

Found in forests on some of the Lesser Antilles, in high elevations and on mountain sides.
    This herbaceous plant grows between 1–2 m high and begins to flower in February. The inflorescence is approximately 30 cm long, and the bracts are yellow above and scarlet below, with small whitish flowers within. The fruit is a blue berry. The leaves are about 0.6–1 m long and about 20–30 cm broad.

Heliconia caribaea Lam.*

Local name: 'Balisier' (Dominica, Fr. West Indies).

Found in the West Indies in middle to upper elevations, growing in moist shady areas along ravines and in thickets.
    A tall herbaceous plant which measures 2–4 m high. The leaves are up to 2 m long. The inflorescence is made up of large red bracts and is more compact than *H. bihai*. There may be as many as eight bracts on the stem. There is also a yellow variety of this species. They begin to flower in March.
    The leaves have been commonly used in thatching and for lining the baskets made by the Caribs out of *Ichnosiphon arouma*. It is also used in covering the local bread when baking.

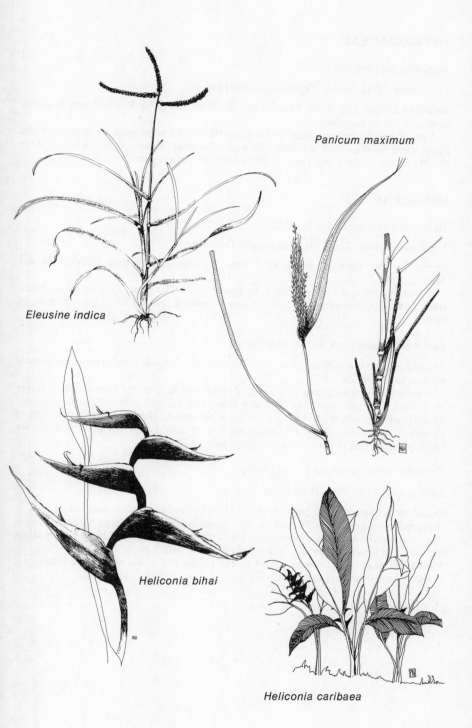

Panicum maximum

Eleusine indica

Heliconia bihai

Heliconia caribaea

115

HYPOXIDACEAE

Hypoxis decumbens L.*

Local name: 'Petit Safran' (Guadeloupe, Martinique).

Found in Central and South America and the West Indies (i). It is common in grassy roadsides and in pastures.

A small erect herbaceous weed with a yellow flower, its outer parts are light green and hairy, while the inner part is yellow with a greenish tinge. Usually there are two or three flowers on the single stem. The capsule is rounded and six-parted.

IRIDACEAE

Belamcanda chinensis (L.) DC.

Local names: Iris; 'Tigré' (Martinique and Guadeloupe).

Originally from China, it is found in some gardens in Dominica, Guadeloupe and Martinique (v).

A low perennial, the leaves sheath the flower stem. The flowers are on loose spikes, erect, and are yellow and orange with brown and purple spots. The flower is in parts of three.

Neomarica caerulea (Ker-Gawl) Sprague

Found from Jamaica to Brazil, and also in tropical Africa. A cultivated ornamental in some gardens in Dominica.

A perennial herb with long sword-shaped leaves arising from the ground. These leaves are smooth and light green. The light blue flowers have deeper blue and purple markings on the petals and yellow and brown markings at the throat. There are usually two or three flowers on each stem. These are spectacular, but last only a day. The plant is useful for borders and hedges in preventing soil dispersal.

Trimezia martinicensis (Jacq.) Herb.*

Local name: 'Coco Chat' (Dominica).

Found on grassy areas and pastures in the West Indies and tropical America (i).

A small erect weed about 30 cm high. The perianth is in parts of three, with three calyx-like petals in the shape of a buttercup. The fruit is three-sided and turns brown when mature. The leaves are narrow, and the plant grows from a small corm.

Medicinal uses Corms were grated by the Caribs and mixed with other plants as a charm to ensure success in battles (xix).

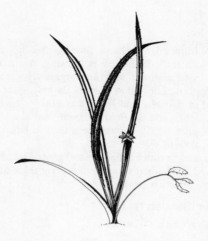

Hypoxis decumbens

Trimezia martinicensis

LILIACEAE

Aloe vera (L.) Burm. f.

Local name: Aloe; 'L'Aloès'.

Native of the Old World tropics (i) the plant is found on beaches and in dry coastal areas.

A herbaceous plant with thick fleshy leaves growing in a rosette from the ground. These are light green, tinged with red, and the margins are slightly spiny. They give a gluey substance when broken. The yellow flowers are on a spike.

Medicinal uses   Used in Barbados and Jamaica as a tea for colds (viii). The leaf is also used for wounds or headaches, and the fluid has been used as a purgative. African tribes made similar uses of the plant (vi). It has also been used in Jamaica to promote appetite, digestion and menstrual flow (iv).

In Marie Galante it is used in rum for diabetes and gas (xvi), and for rubbing the breasts during weaning. It is used on children's fingers to prevent them from sucking the fingers, and also used against eye infection (xvi).

Pleomele sanderiana (Sander) N. E. Br.*

Local names: Lucky Lily; 'Chance' (Dominica).

Native of Western Africa and found in most West Indian gardens.

A slender herbaceous ornamental plant with variegated striped leaves, alternate, sheathing, and lanceolate. About 15 cm long.

MARANTACEAE

Maranta arundinacea L.*

Local names: Arrowroot; 'L'envers' (Fr. West Indies); 'Dictam' (Marie Galante).

Native of Brazil (i) and cultivated in the tropics. Found in semi-shade at lower elevations.

A small plant with ovate leaves which are variegated green and white, and about 15 cm long. The flowers are small, white, with three sepals and a three-lobed corolla.

The rhizomes are used commercially as a form of starch.

Medicinal uses   Used in St. Thomas as a teething aid by pounding the leaves. Also taken as a tea for diarrhoea, and in Barbados the plant was put to similar uses as well (viii). The pounded tuber of any of the varieties was used by the Caribs as a poultice, mixed with wax of a species of large bee. This was formerly used on wounds inflicted by arrows, hence the name 'Arrowroot'. The plant was also used in some magical rites (xix). The crushed rhizomes are used in teas against gas. (xvi).

Ischnosiphon arouma (Aubl.) Koern.*

Local name: 'Larouman'.

Found in the Lesser Antilles and in South America (i). In Dominica it is found along the riverbanks in the northern parts of the island.

A tall plant with large elliptical leaves on erect, long stems. The white flowers are in parts of three with brownish calyxes. The petals are slightly membranous.

The Caribs used the stems when split to weave into baskets. These baskets were usually square with distinctive black and brown colours. They put a layer of leaves into the two woven layers of the basket to make them rainproof. Fans are made in this way — the weave is finer than in the baskets made of *Bambusa vulgaris* and *Passiflora laurifolia*.

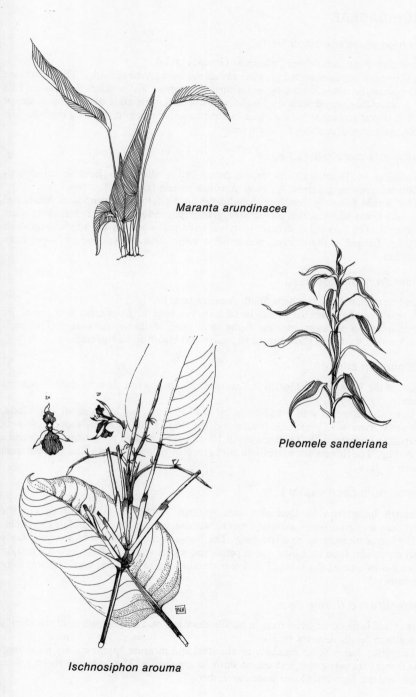

Maranta arundinacea

Pleomele sanderiana

Ischnosiphon arouma

ORCHIDACEAE

Brachionidium sherringii Rolfe*

Sporadic in West Indies from Jamaica to Grenada (xviii).

A very small terrestial orchid growing among the moss on the cliffsides. The flowers are on slender stalks about 4 cm long, subtended by an ovate leaf which is 3 cm long. This leaf is noticeably veined with six parallel veins. The flowers are a deep maroon, almost black in colour and are rather flat, with four petals which taper off into long tendrils. The whole plant measures about 8–10 cm long.

Brassavola cucullata (L.) R. Br.*

Uncommon in Dominica, and found about 240 m above sea level in windswept conditions, growing on trees. A Lesser Antillian species (xviii).

This orchid has very small pseudobulbs (about 1 cm long), a pendulous habit, and unusually linear leaves, which are leathery, grey-green, and have a tiny red marking along the margin. The flower has delicate ivory-coloured petals and sepals, a lighter-coloured ivory lip, fringed at the margin, and it has a sweet scent. It is found in bloom from December.

Epidendrum anceps Jacq.*

Found from Florida to northern South America (xviii).

A small epiphytic orchid found in secondary forests in Dominica. The leaves are approximately 5 cm long, and less fleshy than those of the *E. nocturnum*. The small (1–1.5 cm) flowers are yellow-brown to mauve and bloom from November.

Epidendrum ciliare L.*

Found in the West Indies, Mexico and northern South America. In secondary forests and thickets.

An epiphytic orchid with pseudobulbs up to 15 cm long, and leaves about 20 cm long. The stalk arises with noticeable bracts, and the flowers are on stems about 7 cm long alternating from the stalk. These are about 10–12 cm broad with very narrow petals and a fringed lip. The flowers are white, with faint greenish-yellow markings, and are seen from March to June.

Epidendrum cochleatum L.*

Although uncommon in Dominica, this epiphytic orchid is found throughout the Caribbean area (xviii), on cultivated trees or wooded areas at lower to middle elevations.

The leaves measure up to 20 cm long. The flowers occur in a loose cluster on a fairly short stem, and have thin light-green petals and sepals and a striking deep-maroon lip which has pale-green striations. The flower measures up to 6 cm across and blooms from November.

Epidendrum difforme Jacq.*

This orchid is found on forest trees in middle elevations and ranges from southern Florida to northern South America (xviii).

The fleshy leaves occur regularly on the stem and measure approximately 6 cm long. The flowers are pale green, and almost shiny in appearance. They measure about 1.5 cm across and are found in bloom from November.

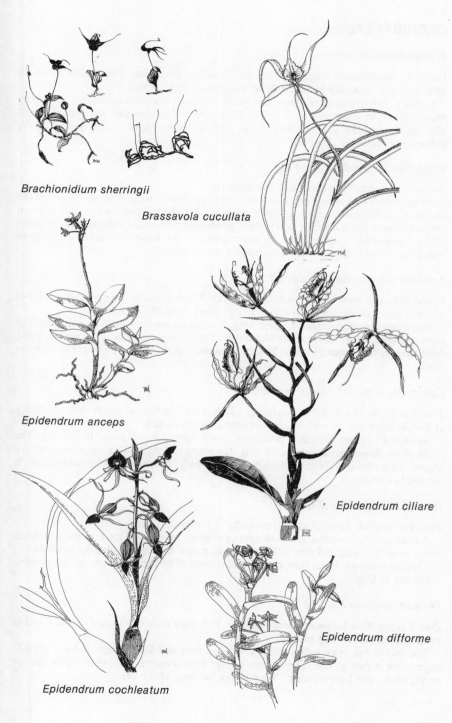

Brachionidium sherringii

Brassavola cucullata

Epidendrum anceps

Epidendrum ciliare

Epidendrum difforme

Epidendrum cochleatum

121

ORCHIDACEAE (continued)

Epidendrum dendrobioides Thunb.*

Found in Guadeloupe, Guyana, Brazil (xviii) and in Dominica. This orchid is rare in Dominica and is usually located only on summits of mountains.

It has a rubbery texture (due perhaps to extreme weather conditions) and shiny appearance. The inflorescence is yellow-green, and the flowers (not shown) are green. The leaves measure about 6–9 cm long, and the boat-shaped yellow bracts of the inflorescence measure about 1 cm long.

Epidendrum jamaicensis Lindl.*

Found in Jamaica, Hispaniola and Dominica (i).

An epiphyte, with almost pendulous branches. The leaves occur towards the end of the branches and measure 8 cm long. The flowers are in a cluster, and are pale yellow-green with a recurved lip, and have a faintly shiny appearance. There are variations in the shape of this lip depending upon where the plant grows, as the small illustration on the right indicates. (Possibly this sketch shows a more recent bloom).

Epidendrum nocturnum Jacq.*

Found from southern Florida to northern South America (xviii), this epiphytic orchid is found at lower elevations up to 650 m in cultivated areas of Dominica.

The stems are approximately 30 cm long. The leaves are stout, and in a cluster. The ivory-coloured flower blooms from November. This flower measures 10 cm across. Slight differences in the formation of the lip may be noticed, depending upon where the plant is found.

Epidendrum pallidiflorum Hook.*

Found in Cuba, Puerto Rico, Guadeloupe, Dominica, Martinique and St. Vincent (xviii), at middle elevations in semi-shade. Uncommon in Dominica.

An orchid with no noticeable pseudobulbs, and narrow leaves about 10 cm long. The stalks of the flowers are covered with long thin bracts and the flowers occur in a terminal cluster. Each flower is about 5 cm long and white with faint magenta stripes on the lip. It blooms in October.

Isochilus linearis (Jacq.) R. Br.*

Found in tropical America and occasionally in Dominica (xviii).

A small epiphytic orchid, found in upper elevations growing on trees, often in montane forest areas in the central part of Dominica. Common throughout tropical America.

This has narrowly linear leaves, and small clusters of bell-shaped, light-purple flowers. It blooms in May.

Malaxis spicata Sw.*

Found in the West Indies and occasionally in Dominica (xviii), this small terrestial orchid most often occurs in rain forests.

The orchid has ovate leaves which are 17 cm long and shiny green. It has a lily-like appearance at first glance. The flowers emerge from a central stalk in a cluster and are small, white, and have no scent. They bloom for most of the year.

Epidendrum dendrobioides

Epidendrum jamaicensis

Epidendrum nocturnum

Epidendrum pallidiflorum

Isochilus linearis

Malaxis spicata

ORCHIDACEAE (continued)

Maxillaria coccinea (Jacq.) L. O. Wms.*

Found in the West Indies, Venezuela, and Columbia (xviii), in rain forests.

A small epiphytic orchid with pseudobulbs and oblong-linear leaves. The scarlet flowers are on single small stems. They grow in clusters and bloom from March to May.

Oncidium jacquinarium Gar. & Stacey.*

Local name: Bee Orchid.

Found in Martinique and in Dominica (xviii) in secondary forests, in semi-shade conditions, growing on trees.

Its pseudobulbs are 15 cm long and the leaves up to 30 cm. The inflorescences are pendulous sprays while the orchid flowers are brown and yellow with brown dots. Many are on one stalk. The little flowers have no scent. Because of the very long sprays of flowers, the plant is seen to best advantage on a fairly tall tree. It blooms in May.

Polystachya sp.*

This epiphytic orchid is pantropical (xviii), common in Dominica, and found at middle to lower elevations.

The plant has very small pseudobulbs, light green leaves, and yellow stalks bearing small insignificant green flowers, which bloom throughout the year.

Spiranthes sp.*

Found throughout tropical America (xviii), in rain forests.

A small terrestial orchid. The white insignificant flowers are on a single stem.

PALMAE

Acrocomia aculeata (Jacq.) Lodd.

Local name: 'GruGru'.

A palm found at lower and drier elevations, in both Dominica and Martinique.

This palm is about 7–8 m tall. It has a fairly sturdy trunk, armed with spines toward the upper part of the tree, which emerge from the rings on the trunk. The leaves are finely pinnate and bushy in appearance. The fruits are roundish, hard on the surface, but with a whitish edible pulp, which is often eaten by children.

Euterpe dominicana L. H. Bailey*

Local name: 'Palmiste'.

Endemic to Dominica in forested areas at middle elevations, usually in the northern part of the island.

This palm grows up to 15 m tall with many leaves. The pinnae are drooping in habit. It is this species which has a terminal bud used in salads and, because of its crisp texture and delicious flavour, is considered a delicacy.

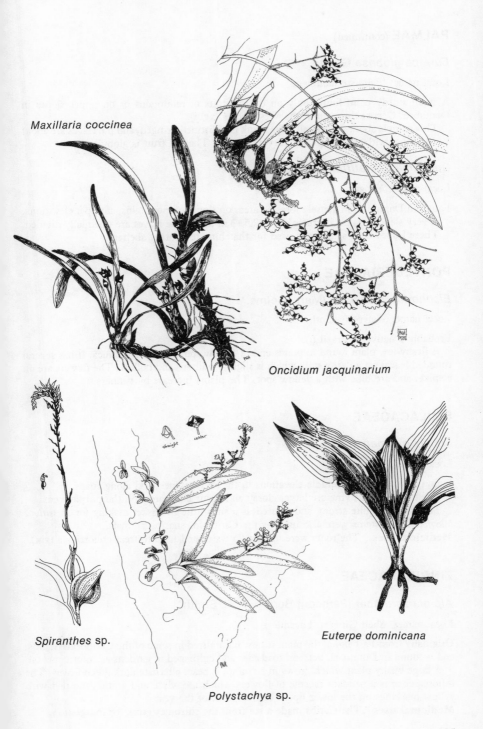

Maxillaria coccinea

Oncidium jacquinarium

Spiranthes sp.

Polystachya sp.

Euterpe dominicana

125

PALMAE (continued)

Euterpe globosa Gaertn.

Local name: 'Palmiste montagne'.

A small sturdy palm found only on the summits of mountains or on upper slopes in Dominica. Occurs throughout the West Indies (xviii).
   This palm is about 2–7 m tall. The trunk is indistinctly ringed, and the leaves are about 1–2 m long, with stiffly upward curving pinnae. The red fruit is globular.

Geonoma pinnatifrons Willd.*

Local name: 'Yanga'.

Found in Dominica, in densely forested areas on slopes of mountains, at upper elevations.
   A very slender palm, it grows about 2.5–3 m high. The leaves are unequally divided. These leaves were sometimes used for thatching temporary shelters.

PONTEDERIACEAE

Eichhornia crassipes (Mart.) Solms.*

Local name: Water Hyacinth.

Probably a native of Brazil (i).
   A fleshy low plant found in ponds and in Freshwater Lake in Dominica. It has several rounded leaves on a fleshy stalk which is slightly bulbous and hollow. The flowers are on a spike, and are blue with a yellow spot. The plants increase by runners.

SMILACACEAE

Smilax guianensis Vitm.*

Local name: 'Bamboche'.

Found in Dominica at middle elevations in thickets, ravines, and along roads.
   The leaves of this vine are large, glossy and occur alternately on the purple stem.
   This stem is quite strong and is used as a substitute in basket making for *Passiflora laurifolia*. The stems were also used by the Caribs for snaring animals.
Medicinal uses   The roots were infused in water and used to treat gonorrhoea (xix).

ZINGIBERACEAE

Alpinia zerumbet (Persoon) Burtt & P. R. Smith*

Local names: Shell Ginger; 'Lavande'.

Originally from East India, the plant is now naturalized in some of the Caribbean islands, and is found in Dominica, both on roadsides and cultivated in gardens as an ornamental.
   A large bushy plant which grows in a clump, it bears alternate dark green leaves. The inflorescence is a pendant raceme of flowers which are white and scarlet, mottled with yellow markings on the lower lip. It blooms most of the year.
Medicinal uses   The Caribs made a tea from the rhizomes (xix), for indigestion.

Geonoma pinnatifrons

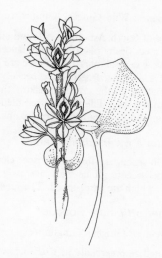

Eichhornia crassipes

Smilax guianensis

Alpinia zerumbet

Hedychium coronarium Koenig.*

Local names: Wild Ginger, 'Gingimbre Doulens', 'Canne d'Eau', 'Canne Rivière'.

A native of South Asia (i), it is found along riverbeds, in middle elevations.

This erect fleshy plant has large lanceolate-elliptic leaves. The flowers are white, with pale yellow central markings. There are usually two or three borne from the calyxes, and they have a strong fragrance, especially in the evening. The rhizomatous roots form a dense matting in the ground, making them difficult to remove once established.

Medicinal uses   The Caribs used the leaves of the plant in a tea for fever (xix), and the root was used in a tea for internal pains.

Renealmia pyramidalis (Lam.)

Found in secondary forests at upper elevations.

An herbaceous plant with narrowly elliptic leaves. It has racemes of small white flowers, which are three-lobed. The green or brown fruits are ovoid. It blooms in March.

Zingiber officinale Roscoe*

Local names: Ginger; 'Gingimbre'.

Cultivated commercially in Dominica.

A small plant about 60 cm high with narrow leaves and small yellow flowers, subtended by red bracts. The rhizome is the commercial ginger. It blooms in December.

Medicinal uses   It is often made into a tea as a digestive stimulant. It was also used for gout (iv) and in West Africa for rheumatic pains, toothache and catarrh (iv).

Hedychium coronarium

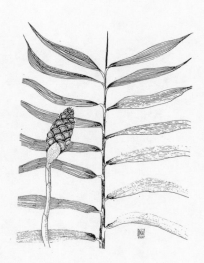

Zingiber officinale

Part 3

MISCELLANEOUS PLANTS

CYATHEACEAE

Cyathea arborea (L.) J. E. Smith.*

Local names: Tree Fern; 'Fougère' (Dominica).

Found in clearings and along roadsides in middle elevations.

An arborescent plant with a single fibrous trunk. The leaves grow from the top and are up to 5 m long, bipinnate, frond-like and arranged spirally.

The trunk is widely used as a plant container, or support for orchids and other epiphytic plants.

There are three species of *Cyathea* in Dominica. One species, *C. imrayana*, has spines on the trunk and leaf stems. When walking in steep areas care must be taken to avoid grasping the stem. The other species, *C. tenera*, is uncommon. The genera *Hemitelia* (two spp.) and *Alsophila* (one sp.) are in this family, and are similar in appearance. (xviii).

LYCOPODIACEAE

Lycopodium cernuum L.*

Local name: 'Cabane de la Vièrge' (Dominica).

Found on banks and cliffs at upper and middle elevations in Dominica, growing in the red clay areas.

A distinctive small fern-like plant, it looks like a small Christmas tree, and is related to *Selaginella* sp. It is light green, furry, and has a shallow creeping root system. The plant will grow to about 30 cm high.

It is used in decorations, especially for religious occasions.

Medicinal uses It produces a bitter tea used for fevers (x).

PODOCARPACEAE

Podocarpus coriaceus L. C. Rich*

Local names: 'Résinier Montagne' (Dominica); 'Laurier Rose' (Martinique and Guadeloupe).

Found in some of the Lesser Antilles islands and on the slopes and upper reaches of some mountains in Dominica.

This is a medium sized tree, with ascending lateral branchlets often appearing on the larger bent branches. The leaves are about 7–8 cm long, almost linear, and acute at the tip. They are whorled on the stems, and are dark green and slightly leathery in texture. The male inflorescence is a spike at the axil of the leaf. The female (not shown) is composed of berries attached to the ends of fleshy wing-like receptacles. It is the only gymnosperm in Dominica.

Cyathea arborea

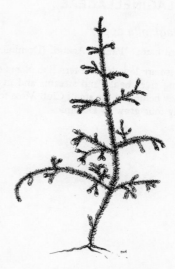

Lycopodium cernuum

Podocarpus coriaceus

PSILOTACEAE

Psilotum nudum (L.) Griseb.*

Pantropical, it is found on trees, stumps, or in the masonry of old buildings.

This epiphytic plant has hanging stems, about 120 cm long, with a matted appearance. The slender two-parted stems eventually divide again. These are all four-angled, and the tiny leaflets are alternate. Green three-lobed sporangia occur alternately on the axillary branches.

This family is generally considered to be the most primitive of the vascular plants.

SELAGINELLACEAE

Selaginella spp.*

Local name: 'Parasol Agouti' (Dominica).

These are low delicate fern-like plants often found on the forest floor. They are common along river banks and streams and in cool shady places at middle to upper elevations. They belong to the small Club Moss family. There are over 600 species world-wide, but only four grow in Dominica.

Psilotum nudum

Selaginella flabellata

Compounds of medicinal interest found in certain plants (vi, vii, viii, ix)

Compound	Plant
Paraglobulin-abrin Phytoalbuminose	*Abrus precatorius*
Fulvine	*Crotalaria fulva*
Aloin Resin Emodin	*Aloe vera*
Tannin Oleo resin Ascorbic acid (Vitamin C)	*Amaranthus species*
Cardole Anacardic acid — nut	*Anacardium occidentale*
Etherel oil Phytosterol Myricyl alcohol Anonol	*Annona muricata*
Anonain — stem and bark	*Annona squamosa*
Ascorbic acid (Vitamin C) — fruit	*Annona squamosa*
Asclepiadin — stem and leaves Vincetoxin — root	*Asclepias curassavica*
Papain Carpain	*Carica papaya*
Capsaicin	*Capsicum frutescens*
Oil of chenopodium	*Chenopodium ambrosoides*
Chrysobarin — seeds	*Cassia occidentalis*
Crescentic acid Chlorogenic acid	*Crescentia cujete*
Vincarosin, Vincristine Vinblastine	*Catharanthus roseus*
Monocrotaline	*Crotalaria incana, Crotalaria retusa*
Hyoscyamine Atropine Scopolamine	*Datura stramonium*
Calcium ethanedioate (Calcium oxalate)	*Diffenbachia seguine*
Saponin — root	*Eryngium foetidum*

137

Compound	Plant
Saponin — root	*Eryngium foetidum*
Saponin	*Guaiacum officinale*
Absinthin	*Compositae Family*
Quercitrin	
Vernonin	
Pelargonin	
Euphorbin	*Euphorbia prostrata*
Narcissine	*Hippeastrum puniceum*
Amaryllin	
Oil of euphorbin	*Hippomane mancinella*
Hydrocyanic acid	*Hura crepitans*
Curcin	*Jatropha curcas*
Rotenone	*Lonchocarpus violasceus*
Mormordocian	*Mormordica charantia*
Carilla	
Nerioderein, Neriodorin	*Nerium oleander*
Karabin	
Oleandrin	
Neriin	
Methyl cinnamate	*Ocimum basilicum*
Calcium ethanedioate (Calcium oxalate)	*Opuntia tuna*
Calcium malate — stem	
Saponin — leaves	*Opuntia sp.*
Flavone — fruit	
Hydrocyanic acid — leaves	*Passiflora foetida*
Ascorbic acid (Vitamin C)	*Peperomia pellucida*
Citric acid	*Plantago major*
Ancubin	
Oil of mustard	*Petiveria alliacea*
Plumbagin	*Plumbago capensis*
Oil of plumbagin	
Ricinoleic acid	*Ruellia tuberosa*
Stearic acid	
Dihydroxystearic acid	
Oleic acid, Linoleic acid	
Ricin	*Ricinus communis*
Tartaric acid	*Spondias mombin*
Tartaric acid	*Tamarindus indicus*
Citric acid	
Ethanoic acid	
Potassium tartrate	

Summary of medicinal plants

Abcesses or boils
Amaranthus dubius
Anthurium palmatum
Bidens pilosa
Carica papaya
Datura stramonium
Hippeastrum puniceum w/*Eryngium*
Hura crepitans

Abortifacients
Carica papaya (seeds)
Crescentia cujete (flesh)
Guaiacum officinale
Mormordica charantia
Nerium oleander
Phyllanthus tenellus
Petiveria alliacea
Plumbago capensis
Wedelia trilobata

Afterbirth
Crescentia cujete
Wedelia trilobata

Analgesic
Tragia volubilis

Aphrodisiacs
Achyranthes indica
Petiveria alliacea

Appetite/digestion
Aloe vera
Alpinia zerumbet
Gouania lupuloides
Zingiber officinale

Asthma
Datura stramonium
Euphorbia hirta
Peperomia spp.
Pilea microphylla

Arthritis
Guaiacum officinale

Baths

Baths for babies
Polygala paniculata

Baths for canoes
Polygala paniculata

Baths for hunting dogs
Hyptis atrorubens
Poivre ginet
Byrsonima spicata

Baths for fatigue
Ceiba pentandra
Citharexylum spinosum
Ocimum basilicum
Pimenta racemosa
Solanum americanum
Urena lobata

Baths for general use
Acacia sp.
Annona muricata
Ceiba pentandra
Citharexylum spinosum
Cymbopogon citratus
Desmodium triflorum
Eleusine indica
Eupatorium triplinerve
Lantana camara
Leonotis nepetifolia
Ocimum basilicum
Pimenta sp.
Pimenta racemosa
Plantago major
Sambucus canadensis
Urena lobata
Wedelia trilobata

Baths for good luck
Aristolochia trilobata
Commelina elegans
Ipomoea pes-caprae
Ipomoea tiliacea
Rubus rosifolius

Baths for parturition
Sauvagesia erecta

Baths for poisoning
Ceiba pentandra
Cymbopogon citratus
Ocimum basilicum

Urena lobata

Baths for prickly heat
Leonotis nepetifolia

Ritual baths
Piper amalago

Baths for skin cleansing
Ageratum conyzoides
Cassia alata
Petiveria alliacea

Biliousness
Capraria biflora
Guaiacum officinale
Ocimum basilicum
Portulaca oleracea

Birth control
Mormordica charantia

Boils (see abscesses)

Bowels
Capraria biflora
Pilea microphylla

Bruises
Crescentia cujete

Cancer
Catharanthus roseus
Mormodica charantia
Peperomia pellucida

Catarrh
Urena lobata
Zingiber officinale

Chest pains
Achyranthes indica

Chills
Lantana involucrata

Cleansing blood
Capraria biflora
Capsicum frutescens

Colds
Achyranthes indica
Ageratum conyzoides
   (with other plants)
Aloe vera
Anacardium occidentale
   (with other plants)
Annona muricata
Begonia hirtella

Begonia macrophylla
Borreria laevis
Cajanus cajan
Capraria biflora
Cassis occidentalis
Cenchrus echinatus
Citrus aurantifolia
Cymbopogon citratus
Episcia mellitiflora
Eryngium foetidum
Justicia pectoralis
Mimosa pudica
Mormordica charantia
Ocimum basilicum
Oldenlandia corymbosa
Passiflora suberosa
Pluchea carolinensis
Peperomia rotundifolia
Petiveria alliacea
Piper amalago
Ruellia tuberosa
Sambucus canadensis
Sauvagesia erecta (with *Polygala paniculata*)
Spondias mombin (as an expectorant in colds)
Waltheria indica
Wedelia trilobata
Urena lobata

Constipation
Opuntia tuna
Piper amalago
Tamarindus indicus

Convulsions
Peperomia pellucida

Coughs
Achyranthes indica
Annona muricata
Antigonon leptopus
Cassia occidentalis
Justicia pectoralis
Leonorus sibiricus
Panicum maximum
Passiflora foetida
Ruellia tuberosa
Spondias mombin
Wedelia trilobata

Cystitis
Ruellia tuberosa
Waltheria indica

Dentifrice
Gouania lupuloides

Depressants
Cannabis sativa
Datura suaveolens

Dermatitis
Cassia alata

Diabetes
Aloe vera
Catharanthus roseus
Cecropia peltata
Mormordica charantia
Peperomia sp.

Diarrhoea
Annona muricata
Cajanus cajan
Capraria biflora
Chrysobalanus icaco
Coccoloba uvifera
Cordia collococca
Eryngium foetidum
Maranta arundinacea
Mimosa pudica
Pilea microphylla
Psidium guajava
Rivina humilis
Rubus rosifolius

Digestion (see appetite)

Diuretic (see urination)
Ageratum conyzoides
Cuscuta americana
Peperomia pellucida

Dropsy
Peperomia pellucida

Dysentery (see diarrhoea)

Dyspepsia
Alpinia zerumbet
Carica papaya
Chrysobalanus icaco
Jatropha gossypifolia
Rubus rosifolius

Earache
Bidens pilosa

Emetic
Asclepias curassavica
Cenchrus echinatus

Enteritis
Ruella tuberosa

Erysipelas
Urena lobata

Eyes
Aloe vera
Bidens pilosa
Capraria biflora
Euphorbia glomifera
Plantago major
Spondias mombin
Stachytarpheta jamaicencis

Fainting
Eryngium foetidum
Peperomia pellucida
Polygala paniculata

Fevers
Achyranthes indica (with
Mimosa pudica)
Annona muricata
Asclepias curassavica
Begonia macrophylla
Begonia hirtella
Capraria biflora
Cenchrus echinatus
Citrus aurantifolia
Cymobopogon citratus
Eupatorium macrophyllum
Euphorbia prostrata
Eryngium foetidum
Hedychium coronarium
Leonotis nepetifolia
Lycopodium cernuum
Morinda citrifolia
Petiveria alliacea
Piper amalago
Stachytarpheta jamaicensis
Thespesia populnea
Tournefortia volubilis

Fits (see fainting)

Gargles/mouthwashes
Cajanus cajan
Gouania lupuloides
Guaiacum officinale
Spondias mombin
Urena lobata

Gas
Aloe vera

Maranta arundinacea
Portulaca oleracea

Gastritis
Urena lobata

Gonorrhoea
Guaiacum officinale
Mangifera indica
Ricinis communis
Smilax guinanensis
Solanum torvum
Spondias mombin
Stachytarpheta jamaicensis

Gout
Guaiacum officinale
Spondias mombin
Zingiber officinale

Gripe
Capsicum frutescens
Opuntia tuna
Jatropha gossypifolia

Hairwash
Opuntia sp.

Headaches
Aloe vera
Clerodendron phillipinum
Cordia retriculata
Morinda citrifolia
Opuntia tuna
Pachystachys coccinea
Petiveria alliacea
Pothomorphe umbellata

Heart
Jatropha curcas
Nerium oleander
Peperomia pellucida

Heat (see baths — prickly heat)

Hypertension (high blood pressure)
Ambrosia hispida
Cassia alata
Lantana involucrata
Peperomia pellucida

Inflammation
Ceiba pentandra
Opuntia sp.
Solanum americanum var. *nodiflorum*

Influenza
Commelina diffusa

Lantana camara
Leonotis nepetifolia
Mangifera indica
Solanum torvum
Stigmaphyllon cordifolium

Intestines
Ruellia tuberosa

Kidneys
Cecropia peltata
Cenchrus echinatus
Cosmos sulphureus
Passiflora foetida
Peperomia pellucida

Labour
Annona squamosa
Annona muricata
Crescentia cujete
Ocimum basilicum
Sauvagesia erecta
Spondias mombin

Laryngitis (see coughs)
Chaptalia nutans

Laxative
Cuscuta americana

Leukemia
Catharanthus roseus

Lice (see mosquitoes)

Malaria
Plantago major

Marasmus
Peperomia pellucida

Measles
Phyllanthus tenellus

Menstruation
Aloe vera
Annona squamosa
Mormordica charantia
Ocimum basilicum
Opuntia tuna
Plantago major
Stigmaphyllon cordifolium
Wedelia trilobata

Mosquitoes
Annona muricata
Annona reticulata
Bixa orellana

142

Ocimum sp.

Mouth (see gargles)

Muscular spasms
Tragia volubilis

Muscular strains
Cosmos sulphureus
Eleusine indica
Ricinus communis
Sida acuta

Neuralgia
Datura stramonium
Ricinus communis

Pleurisy
Eupatorium triplinerve
Urena lobata

Pregnancy (see labour)

Poisoning (snake bite)
Aristolochia trilobata

Poultices
Aloe vera
Amaranthus dubius
Annona reticulata
Bursera simarouba
Caladium bicolor
Chaptalia nutans
Citharexylum fruticosum
Croton flavens
Eleusine indica
Eupatorium triplinerve
Euphorbia prostrata
Guaiacum officinale
Hippeastrum puniceum
Hippobroma longiflora
Hura crepitans
Jatropha curcas
Justicia pectoralis
Kalanchoe pinnata
Morinda citrifolia
Mormordica charantia
Opuntia tuna
Petiveria alliacea
Plantago major
Pluchea carolinensis
Portulaca oleracea
Ricinis communis
Sida acuta
Spondias mombin
Stachytarpheta jamaicensis

Tournefortia volubilis
Urena lobata
Wedelia trilobata

Purges
Aloe vera
Capraria biflora
Cassia alata
Crescentia cujete
Euphorbia prostrata
Jatropha curcas
Mimosa pudica
Mormordica charantia
Ricinis communis
Solanum americanum

Rashes (see baths ‚— prickly heat)

Rheumatic pain
Guaiacum officinale
Hura crepitans
Polygala paniculata
Zingiber officinale

Ringworm
Asclepias curassavica
Carica papaya
Cassia alata

Sedative
Piper amalago

Shock
Annona muricata
Plantago major (with mint and thyme)

Sinus (see asthma)

Skin disorders (see baths
— prickly heat)
Ageratum conyzoides
Cassia alata
Cassia occidentalis
Eupatorium triplinerve
Mormordica charantia
Petiveria alliacea

Smallpox
Cajanus cajan

Snakebite
Aristolochia trilobata

Soporifics
Annona muricata
Cecropia peltata
Cordia collococca

Sores
Hippeastrum puniceum
Plantago major

Spells
Aristolochia sp.
Daphnopsis caribea
Ipomoea pes-caprae
Ocimum micranthum
Rubus rosifolius

Spleen
Annona squamosa

Stimulants
Tragia volubilis
Verbesina alata

Stomach pain
Aristolochia trilobata
Begonia sp.
Canna edulis
Cassia occidentalis
Citrus aurantifolia
Hedychium coronarium
Justicia pectoralis
Phyllanthus sp.
Plantago major
Psidium guajaya

Stricture
Laportea aestuans

Swellings (in legs)
Cassia occidentalis

Syphilis (see gonorrhoea)

Teas
Cooling teas
Ageratum conyzoides
Capraria biflora
Cassia alata
Ceiba pentandra
Commelina diffusa
Eleusine indica
Oxalis corymbosa
Peperomia pellucida
Plantago major
Plumbago capensis
Portulaca oleracea
Ruellia tuberosa
Solanum americanum var. *nodiflorum*
Stachytarpheta cayennensis
Waltheria americana

Hot Teas
Anethum graveolens
Bidens pilosa
Cajanus cajan
Cymbopogon citratus
Eryngium foetidum
Eupatorium triplinerve
Justicia pectoralis
Leonotis nepetifolia
Lycopodium cernuum
Phyllanthus sp.
Plantago major
Psidium guajava
Waltheria americana

Teething
Maranta arundinacea
Stachytarpheta jamaicensis

Tension
Chaptalia nutans

Throat (see coughs)

Tongue (see gargles)
Solanum racemosum

Toothache
Anacardium occidentale
Dacryodes excelsa
Syzgium aromaticum
Tabernaemontana citrifolia
Zingiber officinale

Tuberculosis
Leonotis nepetifolia

Ulcers
Hippomane mancinella
Jatropha curcas
Opuntia tuna
Plantago major
Spondias mombin

Urination (also see diuretic)
Bidens pilosa
Euphorbia hirta
Mimosa pudica
Urena lobata

Vomiting
Centhrus echinatus

Warts
Anacardium occidentale
Carica papaya

Whooping cough
Mimosa pudica

Wind (flatulence)
Alpinia zerumbet
Eupatorium macrophyllum
Plantago major

Worms
Ambrosia hispida
Annona muricata
Asclepias curassavica

Carica papaya
Chenopodium ambrosoides
Leonotis nepetifolia
Mangifera indica
Mormordica charantia
Ricinus communis
Stachytarpheta jamaicensis
Psidium guajaya
Portulaca oleracea

Yaws
Anthurium palmatum

Plants used for fodder

These may be used for rabbits, guinea pigs, or other small livestock.
Amaranthus dubius   4
Bidens pilosa   30
Commelina spp.   106
Emelia spp.   32
Ipomoea tiliaceae   36
Piper amalago   72
Solanum torvum   88

Sources of nectar for honeybees (xxi)

Borreria laevis   80
Commelina diffusa   from morning until plant closes around 11.00 p.m.   106
Cordia curassavica   early morning   20
Cordia laevigata   morning   20
Lantana camara   during the day   96
Leonurus sibiricus   52
Myrcia splendens   64
Mangifera indica   6
Mimosa pudica   from 7.00 a.m. until flower wilts around 10.00 a.m.   62
Mormordica charantia   nectar and pollen from male flowers   38
Petiveria alliacea   early in the morning   70
Sida acuta   a favourite source of nectar for bees — usually between 10.00 a.m. and 2.00 p.m.   58
Stachytarpheta jamaicensis   around noon.   96

Glossary

acuminate Tapering to a point.

adnate Grown attached.

alternate Arranged successively on opposite sides of the stem, or at definite angles. Neither opposite nor whorled.

axil The angle between the upper leaf surface and the stem to which it is attached.

bipinnate When the leaflets of a compound leaf are themselves divided.

bract Modified leaves, usually just beneath the flower.

calyx The group of sepals. They form the outermost part of the flower and are usually green and leaf-like.

campanulate Bell-shaped.

capsule Dry, dehiscent fruit.

compound (leaf) Divided into leaflets.

corolla The group of petals.

cordate Heart-shaped.

corona Appendages found usually, if present, on inner side of the corolla.

crenate Having rounded marginal teeth.

cyme Broad, flattened inflorescence.

drupe Succulent fruit containing a single stone. The stone is part of the wall of the fruit.

epiphyte A plant growing on another plant, but not taking nutrients or water from it.

entire Margin smooth; not toothed.

inflorescence The arrangment of flowers on the stem.

lanceolate Spear-shaped.

lobed Leaves divided, but not into separate leaflets.

lyrate Leaf divided into several lobes with the smallest at the base.

obovate Inversely ovate; with broader end at apex of leaf.

opposite In pairs, attached at same level on opposite sides of the stem.

ovate Egg-shaped.

palmate Lobed to resemble a hand with open fingers.

panicle Compound raceme.

peduncle The flower stalk on which a single flower is borne, or which forms the main stalk of an inflorescence.

peltate Shield-shaped; with petiole joining on the underside to the centre of the leaf.

perianth The floral envelopes: the calyx, corolla, or both.

petiole The leaf-stalk.

pinnate Leaflets of a compound leaf, arranged on each side of a common rachis.

raceme An unbranched inflorescence; each flower has its own stalk attaching it to the main axis.

rhomboid The shape of a diamond in a pack of cards.

scale A papery structure; often a degenerate leaf.

sessile Without a stalk.

spatulate Spoon-shaped.

spike An inflorescence with sessile flowers arising from the main axis.

spur A slender projection from some part of the flower.

stipule One of a pair of leaf-like lateral appendages at the leaf base.

umbel A form of inflorescence in which the flower-stalks of the inflorescence arise from a common centre.

whorl Ring of like parts, such as leaves or petals, around a point on an axis.

Reference list

(i) Adams, C. D., *Flowering plants of Jamaica*, (University of the West Indies, Jamaica, Robt. MacLehose & Co., Ltd., Glasgow, 1972, 848 pp.)

(ii) Adams, C. D., Kasasian, L., Seeyave J., *Common weeds of the West Indies*, (University of the West Indies, Trinidad, 1968, 139 pp.)

(iii) *ARNOLDIA* (Mar./Apr. 1975)

(iv) Asprey, C. F., Thornton, P., *W.I. Medical Journals*(Dec. 1953, Mar. 1954, University of W.I. Jamaica.) Vol. 3, No. 1.

(v) Bayley, A. Dr.

(vi) Bayley, I., 'The Bush Teas of Barbados', *Journal of Barbados Museum and Historical Society* Vol. 16, 103. (1949).

(vii) Bannochie, I. Notes on Bush Teas. (Unpublished Notes.)

(viii) Bannochie, I., Notes on Bush Teas. (Unpublished Notes.)

(ix) Bannochie, I., Teas of Barbados, (Unpublished Notes.)

(x) Dunton, D. 'Rapport sur la Médecine Populaire à la Montagne du Vauclin (Martinique).' (Unpublished paper, University of Montreal, 1977.)

(xi) Duss, R. P., *Flore Phaneogramique des Antilles Francais (Martinique et Guadeloupe)* (Macon, Protat Freres, Imprimeurs, 1897), Reédite, par la *Societe de Distribution et de Culture*, (Martinique, 1972) 2 Vols. 656 pp.

(xii) Fournet, J., *Flore illustrée des phanérogames de Guadeloupe et de Martinique.* (Institut National de la Recherche Agronomique, Paris, 1978, 1654 pp.)

(xiii) Gooding, E. G. B., *The plant communities of Barbados* (Printed at Government Printing Office, Bridgetown, Barbados for Ministry of Education, Barbados, September 1974, 243 pp.)

(xiv) Graf, H. B., *Exotica*, (Roehrs Co., N. J., U.S.A., 1970 Edition, 1833 pp.)

(xv) Graham, V. E., *Tropical wildflowers*, (Hulton Educational Publications, London, 1963, 200 pp.)

(xvi) Grandguillotte, M., *Traditions et Art Populaires de Marie Galante* — Numero 1. 'Les plantes médicinales'. (Murat, 22 juin - 2 juillet 1978, 15 pp.)
and Index Espèces/Indications. (Unpublished notes, 20 mai 1979, 14 pp.)

(xvii) Grisebach, A. H. R., *Flora of the British West Indian Islands.* (J. Cramer, Weinheim, Wheldon & Wesley Ltd., Herts. and Hafner Publishing Co., N.Y., Reprint 1963, 789 pp.)

(xviii) Hodge W. S., 'Flora of Dominica', *Lloydia* Vol. 17, Nos. 1, 2, 3 (June 1954, 238 pp.)

(xix) Hodge W. S., Taylor, D., 'Ethnobotany of the Island Caribs of Dominica', *Webbia*, Vol. XII, No. 2, (1957, pp. 513-644.)

(xx) Imray, J., *Useful Woods of Dominica* (1862)

(xxi) Laurence, G. A., 'Common Bee Weeds of Trinidad and Tobago'. *Journal of the Agricultural Society of Trinidad and Tobago.* Vol. LXXVI, (June 1978, pp. 147-162.)

(xxii) Lawrence G. H. M., *Taxonomy of Vascular Plants*, (The MacMillan Co., N.Y., 1951, 823 pp.)

(xxiii) Macmillan, H. F., *Tropical Planting and Gardening*, (Macmillan & Co., Ltd. London, 1962, 560 pp.)

(xxiv) Menninger, E. A., *Flowering Trees of the World*, (Hearthside Press, N.Y., U.S.A., 1962, 336 pp.)

(xxv) Ringel, G., and Wylie, J., God's Work: Appreciation of the environment in Dominica (Unpublished paper, 1978).

(xxvi) Selawry, O. S., Holland, J. F., Wolman, I. V., The Effect of Vincristine (NSC-67574) on Malignant Solid Tumors in Children. Paper presented at Vincristine Symposium in Memphis, Tennessee, (Jan. 27, 1967).

(xxvii) *U.S. Dept. Agriculture, Publication* No. 882, 'Poisonous and Injurious Plants of the Virgin Islands', (Issued April 1962, 97 pp.)

(xxviii) *W.I. Medical Journal* (June 1970 University of W.I., Jamaica).

(xxix) Wadsworth, F. H., Little, E. L. Jr., 'Common Trees of Puerto Rico and the Virgin Islands'. *Agriculture Handbook* No. 249, (U.S. Dept. of Agriculture, July 1964, 548 pp.)

(xxx) Williams, R. O., Williams, R. O. Jr., *Useful and Ornamental Plants in Trinidad and Tobago* (Government Printery, Trinidad, 1969, 335 pp.) Revised Fourth Edition.

Index to French and patois names

Index to English names

African Tulip Tree (*Spathodea campanulata*) 16

Aloe (*Aloe vera*) 118

Angel's Trumpet (*Datura suavolens*) 86

Annatto (*Bixa orellana*) 16

Annual Weed (*Ocimum basilicum*) 52

Anthurium (*Anthurium* spp.) 100

Arrowroot (*Maranta arundinacea*) 118

Avocado (*Persea americana*) 54

Bald Bush (*Leonotis nepetifolia*) 52

Balsa (*Ochroma pyramidale*) 18

Balsam Apple (*Mormordica charantia*) 38

Bamboo (*Bambusa vulgaris*) 112

Bamboo grass (*Arthrostylidium excelsum*) 112

Bay (*Pimenta racemosa*) 64

Bay Tansy (*Ambrosia hispida*) 30

Bee Orchid (*Oncidium jacquinarium*) 124

Begonia (*Begonia* spp.) 14

Belly Ache Bush (*Jatropha gossypifolia*) 42

Bhaji (*Amaranthus dubius*) 4

Birch Gum (*Bursera simarouba*) 20

Bitter Weed (*Chenopodium ambrosoides*) 28

Black-Eyed Susan (*Thunbergia alata*) 4

Black Sage (*Cordia curassavica*) 20

Blackwattle (*Piper amalago*) 72

Bloodberry (*Rivina humilis*) 70

Blood Flower (*Asclepias curassavica*) 12

Bride's Tears (*Antigonon leptopus*) 76

Broom Weed (*Sida acuta*) 58

Buff Coat (*Waltheria indica*) 88

Burr Grass (*Cenchrus echinatus*) 112

Button Weed (*Borreria laevis*) 80

Calabash (*Crescentia cujete*) 14

Caladium (*Caladium bicolor*) 102

Cankerberry (*Solanum bahamense*) 86

Canna Lily (*Canna indica*) 106

Carib Wood (*Sabinea carinalis*) 48

Carpenter's Grass (*Justicia pectoralis*) 2

Cashew (*Anacardium occidentale*) 6

Castor Oil (*Ricinus communis*) 44

Cat's Blood (*Rivina humilis*) 70

Cat's Claw Creeper (*Macfadyena unguis-cati*) 16

Cerasse (*Mormordica charantia*) 38

Cheddah (*Eleusine indica*) 114

Chinaberry (*Melia azedarach*) 62

Christmas Candles (*Cassia alata*) 24

Christmas Hope (*Tecoma stans*) 16

Christmas Vine (*Porana paniculata*) 36

Clove (*Syzgium aromaticum*) 66

Cochineel (*Opuntia tuna*) 22

Coco Plum (*Chrysobalanus icaco*) 78

Colic Weed (*Achyranthes indica*) 4

Coralita (*Antigonon leptopus*) 76

Coral Vine (*Antigonon leptopus*) 76

Crab's Eyes (*Abrus precatorius*) 44

Cupid's Paint Brush (*Emelia Fosbergii*) 32

Cure For All (*Pluchea carolinensis*) 32

Custard Apple (*Annona reticulata*) 8

Deer Callalou (*Phytolacca icosandra*) 70

Dodder Vine (*Cuscuta americana*) 34

Dumb Cane (*Dieffenbachia siguine*) 102

Duppy Basil (*Ocimum basilicum*) 52

Duppy Gun (*Ruellia tuberosa*) 2

Easter Lily (*Hippeastrum puniceum*) 100

Fat Pork (*Chrysobalanus icaco*) 78

Fit Weed (*Eryngium foetidum*) 92

French Weed (*Commelina diffusa*) 106

Garden Balsam (*Justicia pectoralis*) 2

Garlic Weed (*Petiveria alliacea*) 70

Ginger (*Zingiber officinale*) 128

Ginger Thomas (*Tecoma stans*) 16

Goatweed (*Capraria biflora*) 84

Golden Shower (*Macfadyena unguis-cati*) 16

Grass (*Cannabis sativa*) 26

Guava (*Psidium guajava*) 64

Guinea Grass (*Panicum maximum*) 114

Guinea Hen Weed (*Petiveria alliacea*) 70

Gully Root (*Petiveria alliacea*) 70

Haiti Haiti (*Thespesia populnea*) 58

Hog Plum (*Spondias mombin*) 6

Hog Weed (*Boerhavia coccinea*) 66

Horse Poison (*Hippobroma longiflora*) 24

Index to scientific names